Crowdsourced Data Management

Guoliang Li • Jiannan Wang • Yudian Zheng
Ju Fan • Michael J. Franklin

Crowdsourced Data Management

Hybrid Machine-Human Computing

Springer

Guoliang Li
Department of Computer Science
and Technology
Tsinghua University
Beijing, Beijing, China

Yudian Zheng
Twitter Inc.
San Francisco, CA, USA

Michael J. Franklin
Department of Computer Science
University of Chicago
Chicago, IL, USA

Jiannan Wang
School of Computing Science
Simon Fraser University
Burnaby, BC, Canada

Ju Fan
DEKE Lab & School of Information
Renmin University of China
Beijing, Beijing, China

ISBN 978-981-13-4012-3   ISBN 978-981-10-7847-7   (eBook)
https://doi.org/10.1007/978-981-10-7847-7

This Springer imprint is published by the registered company Springer Nature Singapore Pte Ltd.
The registered company address is: 152 Beach Road, #21-01/04 Gateway East, Singapore 189721, Singapore

Preface

Many important data management and analytics tasks, such as entity resolution, sentiment analysis, and image recognition, can be enhanced through the use of human cognitive ability. Crowdsourcing platforms provide an effective way of harnessing the capabilities of people (i.e., the crowd) to process such tasks, and they encourage many real-world applications, such as reCaptcha, ImageNet, ESP game, Foldit, and Waze. Recently, crowdsourced data management has attracted increasing interest from both academia and industry.

This book provides a comprehensive review of crowdsourced data management, including motivation, applications, techniques, and existing systems. The book first introduces an overview of crowdsourcing, including crowdsourcing motivation, background, applications, workflows, and platforms. For example, consider the entity resolution problem, which, given a set of objects, finds the objects that refer to the same entity. Since machine algorithms cannot achieve high quality for this problem, crowdsourcing can be used to improve the quality. A user (called the "requester") first generates some relevant tasks (e.g., asking whether two objects refer to the same entity), configures them (e.g., setting the price, latency), and then posts them onto a crowdsourcing platform. Users (called "workers") that are interested in these tasks can accept and answer them. The requester pays the participating workers for their labor. Crowdsourcing platforms manage the tasks and assign them to the workers.

Then, this book summarizes three important problems in crowdsourced data management: (1) quality control: workers may return noisy or incorrect results, so effective quality-control techniques are required to improve the quality; (2) cost control: the crowd is often not free, and cost control aims to reduce the monetary cost; (3) latency control: human workers can be slow, particularly compared to automated computing time scales, so latency-control techniques are required to control the pace. There have been significant works addressing these three factors for designing crowdsourced tasks, developing crowdsourced data manipulation operators, and optimizing plans consisting of multiple operators.

Next, this book synthesizes a wide spectrum of existing studies on crowdsourced data management, including crowdsourcing models, declarative languages, crowdsourced operators, and crowdsourced optimizations.

Finally, based on this analysis, this book outlines key factors that need to be considered to improve crowdsourced data management.

Beijing, China   Guoliang Li
Burnaby, BC, Canada   Jiannan Wang
San Francisco, CA, USA   Yudian Zheng
Beijing, China   Ju Fan
Chicago, IL, USA   Michael J. Franklin
December 2017

Acknowledgments

We express our deep gratitude to those who have helped us in writing this book. We thank Chengliang Chai, James Pan, and Jianhua Feng for discussing or proofreading the earlier versions of the book. Our thanks also go to Lanlan Chang and Jian Li at Springer for their kind help and patience during the preparation of this book.

We acknowledge the financial support by the 973 Program of China (2015CB358700), the NSF of China (61632016, 61373024, 61602488, 61422205, 61472198), TAL education, Tencent, Huawei, and FDCT/007/2016/AFJ.

Contents

Chapter 1
Introduction

Crowdsourcing has been widely used to enhance many data management and analytics tasks. This chapter introduces the motivation, overview, and research challenges of crowdsourced data management. Section 1.1 provides a motivation of crowdsourcing, and Sect. 1.2 gives a brief overview of crowdsourcing. Finally, Sect. 1.3 introduces research challenges of crowdsourced data management and summarizes existing works.

1.1 Motivation

Computers have changed the world by storing and processing huge amounts of information. However, there are still many machine-hard problems that cannot be completely addressed by automated processes. For example, consider sentiment analysis that aims to identify the sentiment of a sentence or a picture [30, 35, 63], handwriting recognition [17], and image recognition [44, 56, 59]. Although there are some machine-based algorithms for these tasks, better performance is achievable through the use of human cognitive ability. For instance, we can ask a human to identify the sentiment of a message, recognize handwriting, and recognize an image, and the human will usually have higher accuracy than the machine algorithms.

Crowdsourcing provides an effective way to process such machine-hard tasks by utilizing hundreds of thousands of ordinary workers (i.e., the crowd). Fortunately, access to crowd resources has been made easier due to public crowdsourcing platforms, such as Amazon Mechanical Turk (AMT) [1], CrowdFlower [2], and Upwork [3]. As a result, crowdsourcing has attracted significant attention from both industry and academia, and it has become an active area in the human-computer interaction and data management communities [28].

There are many successful applications that utilize crowdsourcing to solve machine-hard tasks, e.g., sentiment analysis, search relevance (checking whether

© Springer Nature Singapore Pte Ltd. 2018
G. Li et al., *Crowdsourced Data Management*,
https://doi.org/10.1007/978-981-10-7847-7_1

a search result is relevant to the query), content moderation (checking whether a website contains adult content), data collection (collecting questionnaires), data categorization (categorizing a picture to a category), and image/audio transcription (transforming an image/audio to text). In particular, a famous application is reCAPTCHA. Many websites ask users to recognize the words from a given image before they are allowed to log into these websites. This process is an example of reCAPTCHA. Von Ahn et al. [51] utilized reCAPTCHA to digitize New York newspapers. First, the newspapers were scanned as images and then split into words. Next, new images with two words were generated – one word selected from the scanned newspapers and the other word generated by machines. Then, these generated images were used as verification codes by asking Internet users to recognize the words from the images. Users must carefully recognize the words in order to pass the verification code, and in this way, Internet users were harnessed to transcribe all the newspapers into text without spending any money. This method achieved an accuracy exceeding 99% and has since been used to transcribe over 440 million words. As another example, ESP game [4] designed a game to ask users to label images. The game selected an image and asked two players to identify objects from the image. If the two players returned the same answer, then they were successful and asked to label the next image. Many players tried to label as many images as possible, and in this way, ESP game utilized the players to label many images. Similarly, ImageNet [12] utilized crowdsourcing to construct an image hierarchy dataset. The program asked humans to categorize an image to a category in a knowledge hierarchy, e.g., given a dog image, users were asked to categorize it to "Animal, Dog, Husky." Finally, Eiben et al. [13] designed a game Foldit that utilized a game-driven crowdsourcing method to enhance a computationally designed enzyme. In the game, humans used their stereo-sensory capacity to solve the structure of an enzyme. Foldit was able to successfully solve the riddle of the HIV enzyme within 3 weeks.

In summary, crowdsourcing has become an effective way to address real-world applications.

1.2 Crowdsourcing Overview

We introduce the crowdsourcing workflow. There are three parties involved in crowdsourcing – requesters, the platform, and workers. Requesters have machine-hard jobs to be crowdsourced. A requester generates tasks and uploads these tasks to a crowdsourcing platform, e.g., AMT. Next, the crowdsourcing platform publishes the tasks. Crowd workers who are willing to perform such tasks (typically for pay or some other reward) accept the tasks, answer them, and submit the answers back to the platform. The platform collects the answers and reports them to the requester.

Next, we take entity resolution as an example to show the details of the crowdsourcing workflow. Given a set of objects, entity resolution aims to identify the object pairs that refer to the same entity. For example, given a set of products from Amazon and eBay, the aim is to find all product pairs that refer to the same real-world product, e.g., "Apple iPhone 6s Plus 128 GB" from Amazon and "iPhone 6 s 5.5 inch 128 GB" from eBay should refer to the same product; but "Apple iPhone 6s Plus 128 GB" and "iPhone 6 s 4.7 inch 128 GB" refer to different products. Although this problem has been studied for decades, traditional algorithms are still far from perfect [50, 52, 54, 54, 55], and crowdsourcing can be utilized to address this problem by harnessing the crowd's ability.

Requester A requester needs to design the tasks, set up the tasks, publish the tasks to crowdsourcing platforms, and monitor the tasks. For entity resolution, task design aims to design effective task types. For example, for every pair of objects, devising a YES/NO question that asks workers to indicate whether the two objects refer to the same entity is a task design. Another possible task design is to provide workers with multiple objects and ask the workers to find the pair of objects that refer to the same entity. Task setting sets the properties of tasks, e.g., deciding the price of each task, setting the time constraint for each task (the task will time out if workers do not answer the task within the constraint), and choosing quality-control methods (for improving the result quality). Task monitoring aims to help requesters monitor the tasks and collect the results from the crowdsourcing platform.

Tasks can be broadly classified into macro-tasks and micro-tasks, where the former is a complicated task (e.g., developing a mobile-phone app or designing a logo) and may take several hours/days to complete, while the latter is an easy task (e.g., labeling an image) and usually takes a couple of seconds or minutes for a worker to answer. There are several strategies to attract workers to answer tasks, including monetary rewards, interest-based, and volunteering. In this book, we focus on micro-tasks (called tasks in this book) and monetary rewards (e.g., requesters need to pay the workers who help them complete the tasks).

Worker Workers just need to browse the tasks on the crowdsourcing platform, select interesting tasks, and submit the answers to the platform. Workers aim to get money from requesters by helping them answer questions. Crowd workers have different characteristics from computers. (1) *Not Free*. Workers need to be paid for answering a task, and the requester usually has a budget to control the *cost*. (2) *Error Prone and Diverse*. Workers may return noisy results, and it is important to tolerate noise and improve the *quality*. Because workers have various background knowledge, and hence different accuracies for answering different tasks, workers' characteristics need to be captured in order to achieve high *quality*. (3) *Dynamic*. Workers are not always online to answer tasks, and it is important to control the *latency*.

Crowdsourcing Platform Crowdsourcing platforms provide a system such that requesters can deploy tasks and workers can answer them. Requesters need to pay a crowdsourcing platform (e.g., 5% of the money paid to workers) in order to use it. Platforms help requesters manage tasks and collect answers. Some workers may give low-quality results, and platforms provide quality-control strategies to help requesters improve the results quality. The first strategy is a *qualification test*. The requester provides several *golden tasks* with known ground truth. Before workers answer a requester's task, the workers must pass the qualification test, like an exam. The second strategy is a *hidden test*. The requester can mix the golden tasks into the normal tasks. If the workers give low-quality answers on the hidden golden tasks, the requester can ask the platform to block the workers. The difference between qualification tests and hidden tests is that the requester needs to pay workers for the golden tasks in hidden tests but does not need to pay workers for qualification tests.

1.3   Crowdsourced Data Management

This book focuses on crowdsourced data management that utilizes crowdsourcing to improve data management and analytics problems, such as query processing [19, 37, 38, 43, 50, 52, 54, 55, 57, 59], data cleaning [41, 53], data integration [23, 26, 29, 58], and knowledge construction [5, 8]. Next, we introduce the challenges that need to be addressed in crowdsourced data management.

Major Challenges Traditional data management systems usually focus on optimizing one optimization goal – reducing the latency. However, due to the different features of workers (not free, error prone and diverse, dynamic), there are three optimization goals in crowdsourcing as shown in Fig. 1.1.

(1) Improving Quality Crowd workers may return relatively low-quality results or even noise. For example, a malicious worker may intentionally give wrong answers. Workers may have different levels of expertise, and an untrained worker may be incapable of accomplishing certain tasks. To achieve high quality, crowd errors should be tolerated, and high-quality results should be inferred from noisy answers. The first step of quality control is to characterize a worker's quality (called worker modeling). For example, a worker's quality can be simply modeled as a probability, e.g., 0.8, i.e., the worker has a probability of 0.8 to correctly answer a task. The probability can be computed based on a qualification test, a hidden test or more sophisticated algorithms. If a worker answers ten tasks, among which eight tasks are correct, then her quality is 0.8. More sophisticated models will be introduced in Chap. 3. Then, based on the quality model of workers, there are several strategies to improve the quality in crowdsourcing. First, low-quality workers can be eliminated (called worker elimination). For example, workers whose quality is below 0.6 can be blocked. Second, a task can be assigned to multiple workers, and the truth (the correct answer) of the task can be inferred by aggregating workers' results (called truth inference). For example, each task can be assigned to five workers, and then majority voting can be used to aggregate the answer. Third, a task can be assigned to appropriate workers (called task assignment) who are good at such tasks. More quality-control methods will be discussed in Chap. 3.

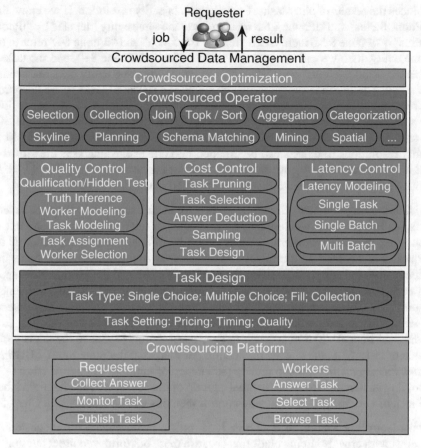

Fig. 1.1 Overview of crowdsourced data management

(2) Reducing Cost The crowd is not free, and if there are large numbers of tasks, crowdsourcing can be expensive. For example, in entity resolution, if there are 10,000 objects, there will be about 50 million possible pairs. Even if the price per pair is 1 cent, the total cost is half a million dollars. There are several effective cost-control techniques. The first is pruning, which first uses machine algorithms to remove some unnecessary tasks and then utilizes the crowd to answer only the necessary tasks. In the entity resolution example, highly dissimilar pairs (e.g., iPhone vs Samsung) can be pruned by using machine algorithms, and these tasks do not need to be crowdsourced. The second is task selection, which prioritizes the tasks and decides which tasks to crowdsource first. For example, suppose an entity resolution task is assigned to five workers iteratively. If three workers return that the two objects in the task refer to the same entity, the fourth and fifth workers do not need to be asked, thus saving two tasks. The third is answer deduction, which crowdsources a subset of tasks, and based on the answers collected from the crowd deduces the results of other tasks. For example, in entity resolution, if we know that "iPhone 8 Plus" and "iPhone 8 5.5 inch" refer to the same entity (denoted by "iPhone 8 Plus" = "iPhone 8 5.5 inch"), and "iPhone 8 5.5 inch" and "iPhone 8+" refer to the same entity, then we can infer that "iPhone 8 Plus" = "iPhone 8+", and the task of whether "iPhone 8 Plus" and "iPhone 8+" refer to the same entity does not need to be asked. The fourth is sampling, which samples a subset of tasks to crowdsource. For example, suppose the requester wants to know the number of distinct entities in a dataset. Sampling can be used to compute an approximate number. The fifth is task design, which designs a better user interface for the tasks. For example, in entity resolution, 10 objects can be grouped into one task which asks the crowd to identify the matching objects, rather than asking the crowd to compare $\frac{10 \cdot 9}{2} = 45$ pairs. The cost-control methods will be discussed in Chap. 4.

(3) Controlling Latency   Crowd answers may incur excessive latency for several reasons. For example, workers may be distracted or unavailable, tasks may not be appealing to enough workers, or tasks might be difficult for most workers. If the requester has a time constraint, it is important to control the latency. Note that latency does not simply depend on the number of tasks and the average time spent on each task, because crowd workers perform tasks in parallel. Existing latency-control techniques can be classified into three categories. (1) *Single-task latency control* aims to reduce the latency of one task (e.g., the latency of labeling each individual image). (2) *Single-batch latency control* aims to reduce the latency of a batch of tasks (e.g., the latency of labeling ten images at the same time). (3) *Multi-batch latency control* aims to reduce the latency of multiple batches of tasks (e.g., adopting an iterative workflow to label a group of images where each iteration labels a batch of two images). The latency-control methods will be discussed in Chap. 5.

Crowdsourcing Database Systems   It is rather inconvenient for requesters to interact with crowdsourcing platforms because these platforms require requesters to be familiar with the user interface to deploy tasks and design tasks, set parameters (e.g., the number of workers for each task), and even write codes (e.g., designing custom task types). To encapsulate the complicated process of interacting with the crowd requires designing crowdsourcing database systems that enable users to utilize declarative queries (e.g., SQL) to set up and deploy tasks. Moreover, most crowdsourcing platforms do not provide effective techniques for quality control, cost control, and latency control. This absence calls for a system to provide effective techniques to address these problems. In addition, crowdsourcing database systems can make databases more powerful by harnessing human abilities to search and collect results. For example, a requester searches for "US" companies in a database, but the data in the database uses "USA" and "America." Thus a traditional database retrieves no results, while the crowdsourcing database can easily find the answers. Recently, several crowdsourcing database systems, such as CrowdDB [18], Qurk [34], Deco [40], CrowdOP [16], and CDB [27], have been designed. Crowd-sourcing database systems focus on devising declarative query languages (e.g., extending SQL to support crowdsourcing operations), supporting query operators, designing query models, and devising query optimization techniques to balance cost (cheap), latency (fast), and quality (good). The details of database systems will be discussed in Chap. 6.

Crowdsourced Data Management Operators   Crowdsourced database operators are very important to designing crowdsourcing database systems. For example, entity resolution can use a crowdsourced join operator to find objects referring to the same entity. In data extraction, one needs to use a crowdsourced selection operator to select relevant data. In subjective comparison scenarios, a crowdsourced sort operator is used to rank the results. All database operators have been studied in crowdsourcing scenarios. In particular, *crowdsourced selection* aims to utilize the crowd to select some objects from a set of objects that satisfy a filtering condition [37, 38, 43, 59], e.g., selecting photos containing a mountain and a lake. *Crowdsourced collection* aims to ask the crowd to collect a set of objects [41, 48], e.g., collecting active NBA players born in the USA. *Crowdsourced join* is similar to crowdsourced entity resolution that utilizes the crowd to find objects referring to the same entity [19, 50, 52, 54, 55, 57]. *Crowdsourced top-k/sort* harnesses the crowd to sort/rank a set of objects [10, 11, 14, 21, 21, 25, 42, 60], e.g., ranking the reviews of a product on Amazon. *Crowdsourced categorization* aims to utilize the crowd to categorize an object to a predefined category [39], e.g., categorizing a photo to a category in the ImageNet knowledge hierarchy. *Crowdsourced aggregation* aims to utilize the crowd to support aggregation queries (e.g., count and median) [11, 21, 22, 33, 49], e.g., asking the crowd to count the number of people on a beach. *Crowdsourced skyline* [20, 31, 32] harnesses the crowd to find an object that is not dominated by other objects in terms of any attribute. For example, finding a hotel by considering multiple factors, e.g., convenient transportation, low price, high rating, clean, etc., the crowdsourced skyline operator can ask the crowd to find appropriate hotels that are not worse than any other hotels in each factor. *Crowdsourced planning* aims to design a sequence of actions from an initial state to reach a goal [24, 45, 46, 62, 64]. For example, a student wants to plan the sequence of courses to take in a semester, in order to obtain a solid knowledge of a specific domain, by asking upperclassman. *Crowdsourced schema matching* aims to utilize the crowd to find mappings of attributes in two tables [15, 36, 61], e.g., finding the mappings between an Amazon product table and an eBay product table. *Crowdsourced mining* tries to learn and observe significant patterns based on crowd answers [6–9]. For example, one can discover the association rules such that *garlic can be used to treat flu* by asking many workers. *Spatial crowdsourcing* aims to ask the crowd to answer spatial tasks [47], e.g., checking whether there is a parking spot and taking a photo of a restaurant to check how long the wait is for dining. These operators are fundamental to designing crowdsourcing systems and for supporting crowdsourcing applications. How to effectively support crowdsourced operators will be discussed in Chap. 7.

Finally, we summarize existing works in crowdsourcing and provide several research challenges and open problems in Chap. 8.

References

1. Amazon mechanical turk. https://www.mturk.com/
2. Crowdflower. http://www.crowdflower.com
3. Upwork. https://www.upwork.com
4. von Ahn, L., Dabbish, L.: ESP: labeling images with a computer game. In: AAAI, pp. 91–98 (2005)
5. Amsterdamer, Y., Davidson, S., Kukliansky, A., Milo, T., Novgorodov, S., Somech, A.: Managing general and individual knowledge in crowd mining applications. In: CIDR (2015)
6. Amsterdamer, Y., Davidson, S.B., Milo, T., Novgorodov, S., Somech, A.: Oassis: query driven crowd mining. In: SIGMOD, pp. 589–600. ACM (2014)
7. Amsterdamer, Y., Davidson, S.B., Milo, T., Novgorodov, S., Somech, A.: Ontology assisted crowd mining. PVLDB 7(13), 1597–1600 (2014)
8. Amsterdamer, Y., Grossman, Y., Milo, T., Senellart, P.: Crowd mining. In: SIGMOD, pp. 241–252. ACM (2013)
9. Amsterdamer, Y., Grossman, Y., Milo, T., Senellart, P.: Crowdminer: Mining association rules from the crowd. PVLDB 6(12), 1250–1253 (2013)
10. Chen, X., Bennett, P.N., Collins-Thompson, K., Horvitz, E.: Pairwise ranking aggregation in a crowdsourced setting. In: WSDM, pp. 193–202 (2013)
11. Davidson, S.B., Khanna, S., Milo, T., Roy, S.: Using the crowd for top-k and group-by queries. In: ICDT, pp. 225–236 (2013)
12. Deng, J., Dong, W., Socher, R., Li, L.J., Li, K., Fei-Fei, L.: ImageNet: A Large-Scale Hierarchical Image Database. In: CVPR (2009)
13. Eiben, C.B., Siegel, J.B., Bale, J.B., Cooper, S., Khatib, F., Shen, B.W., Players, F., Stoddard, B.L., Popovic, Z., Baker, D.: Increased diels-alderase activity through backbone remodeling guided by foldit players. Nature biotechnology 30(2), 190–192 (2012)
14. Eriksson, B.: Learning to top-k search using pairwise comparisons. In: AISTATS, pp. 265–273 (2013)
15. Fan, J., Lu, M., Ooi, B.C., Tan, W.C., Zhang, M.: A hybrid machine-crowdsourcing system for matching web tables. In: ICDE, pp. 976–987. IEEE (2014)
16. Fan, J., Zhang, M., Kok, S., Lu, M., Ooi, B.C.: Crowdop: Query optimization for declarative crowdsourcing systems. IEEE Trans. Knowl. Data Eng. 27(8), 2078–2092 (2015)
17. Fang, Y., Sun, H., Li, G., Zhang, R., Huai, J.: Effective result inference for context-sensitive tasks in crowdsourcing. In: DASFAA, pp. 33–48 (2016)
18. Franklin, M.J., Kossmann, D., Kraska, T., Ramesh, S., Xin, R.: Crowddb: answering queries with crowdsourcing. In: SIGMOD, pp. 61–72 (2011)
19. Gokhale, C., Das, S., Doan, A., Naughton, J.F., Rampalli, N., Shavlik, J.W., Zhu, X.: Corleone: hands-off crowdsourcing for entity matching. In: SIGMOD, pp. 601–612 (2014)
20. Groz, B., Milo, T.: Skyline queries with noisy comparisons. In: PODS, pp. 185–198 (2015)
21. Guo, S., Parameswaran, A.G., Garcia-Molina, H.: So who won?: dynamic max discovery with the crowd. In: SIGMOD, pp. 385–396 (2012)
22. Heikinheimo, H., Ukkonen, A.: The crowd-median algorithm. In: HCOMP (2013)
23. Ipeirotis, P., Provost, F., Wang, J.: Quality management on amazon mechanical turk. In: SIGKDD Workshop, pp. 64–67 (2010)
24. Kaplan, H., Lotosh, I., Milo, T., Novgorodov, S.: Answering planning queries with the crowd. PVLDB 6(9), 697–708 (2013)
25. Khan, A.R., Garcia-Molina, H.: Hybrid strategies for finding the max with the crowd. Tech. rep. (2014)
26. Li, G.: Human-in-the-loop data integration. PVLDB 10(12), 2006–2017 (2017)
27. Li, G., Chai, C., Fan, J., Weng, X., Li, J., Zheng, Y., Li, Y., Yu, X., Zhang, X., Yuan, H.: CDB: optimizing queries with crowd-based selections and joins. In: SIGMOD, pp. 1463–1478 (2017)
28. Li, G., Wang, J., Zheng, Y., Franklin, M.J.: Crowdsourced data management: A survey. TKDE 28(9), 2296–2319 (2016)

29. Liu, Q., Peng, J., Ihler, A.T.: Variational inference for crowdsourcing. In: NIPS, pp. 701–709 (2012)
30. Liu, X., Lu, M., Ooi, B.C., Shen, Y., Wu, S., Zhang, M.: CDAS: A crowdsourcing data analytics system. PVLDB 5(10), 1040–1051 (2012)
31. Lofi, C., Maarry, K.E., Balke, W.: Skyline queries in crowd-enabled databases. In: EDBT, pp. 465–476 (2013)
32. Lofi, C., Maarry, K.E., Balke, W.: Skyline queries over incomplete data - error models for focused crowd-sourcing. In: ER, pp. 298–312 (2013)
33. Marcus, A., Karger, D.R., Madden, S., Miller, R., Oh, S.: Counting with the crowd. PVLDB 6(2), 109–120 (2012)
34. Marcus, A., Wu, E., Madden, S., Miller, R.C.: Crowdsourced databases: Query processing with people. In: CIDR, pp. 211–214 (2011)
35. Mozafari, B., Sarkar, P., Franklin, M., Jordan, M., Madden, S.: Scaling up crowd-sourcing to very large datasets: a case for active learning. PVLDB 8(2), 125–136 (2014)
36. Nguyen, Q.V.H., Nguyen, T.T., Miklós, Z., Aberer, K., Gal, A., Weidlich, M.: Pay-as-you-go reconciliation in schema matching networks. In: ICDE, pp. 220–231. IEEE (2014)
37. Parameswaran, A.G., Boyd, S., Garcia-Molina, H., Gupta, A., Polyzotis, N., Widom, J.: Optimal crowd-powered rating and filtering algorithms. PVLDB 7(9), 685–696 (2014)
38. Parameswaran, A.G., Garcia-Molina, H., Park, H., Polyzotis, N., Ramesh, A., Widom, J.: Crowdscreen: algorithms for filtering data with humans. In: SIGMOD, pp. 361–372 (2012)
39. Parameswaran, A.G., Sarma, A.D., Garcia-Molina, H., Polyzotis, N., Widom, J.: Human-assisted graph search: it's okay to ask questions. PVLDB 4(5), 267–278 (2011)
40. Park, H., Pang, R., Parameswaran, A.G., Garcia-Molina, H., Polyzotis, N., Widom, J.: Deco: A system for declarative crowdsourcing. PVLDB 5(12), 1990–1993 (2012)
41. Park, H., Widom, J.: Crowdfill: collecting structured data from the crowd. In: SIGMOD, pp. 577–588 (2014)
42. Pfeiffer, T., Gao, X.A., Chen, Y., Mao, A., Rand, D.G.: Adaptive polling for information aggregation. In: AAAI (2012)
43. Sarma, A.D., Parameswaran, A.G., Garcia-Molina, H., Halevy, A.Y.: Crowd-powered find algorithms. In: ICDE, pp. 964–975 (2014)
44. Smyth, P., Fayyad, U.M., Btruth from subjective labelling of venus images. In: NIPS, pp. 1085–1092 (1994)
45. Su, H., Zheng, K., Huang, J., Jeung, H., Chen, L., Zhou, X.: Crowdplanner: A crowd-based route recommendation system. In: ICDE, pp. 1144–1155. IEEE (2014)
46. Talamadupula, K., Kambhampati, S., Hu, Y., Nguyen, T.A., Zhuo, H.H.: Herding the crowd: Automated planning for crowdsourced planning. In: HCOMP (2013)
47. To, H., Ghinita, G., Shahabi, C.: A framework for protecting worker location privacy in spatial crowdsourcing. PVLDB 7(10), 919–930 (2014)
48. Trushkowsky, B., Kraska, T., Franklin, M.J., Sarkar, P.: Crowdsourced enumeration queries. In: ICDE, pp. 673–684 (2013)
49. Venetis, P., Garcia-Molina, H., Huang, K., Polyzotis, N.: Max algorithms in crowdsourcing environments. In: WWW, pp. 989–998 (2012)
50. Vesdapunt, N., Bellare, K., Dalvi, N.N.: Crowdsourcing algorithms for entity resolution. PVLDB 7(12), 1071–1082 (2014)
51. Von Ahn, L., Maurer, B., McMillen, C., Abraham, D., Blum, M.: recaptcha: Human-based character recognition via web security measures. Science 321(5895), 1465–1468 (2008)
52. Wang, J., Kraska, T., Franklin, M.J., Feng, J.: CrowdER: crowdsourcing entity resolution. PVLDB 5(11), 1483–1494 (2012)
53. Wang, J., Krishnan, S., Franklin, M.J., Goldberg, K., Kraska, T., Milo, T.: A sample-and-clean framework for fast and accurate query processing on dirty data. In: SIGMOD, pp. 469–480 (2014)
54. Wang, J., Li, G., Kraska, T., Franklin, M.J., Feng, J.: Leveraging transitive relations for crowdsourced joins. In: SIGMOD, pp. 229–240 (2013)

55. Wang, S., Xiao, X., Lee, C.: Crowd-based deduplication: An adaptive approach. In: SIGMOD, pp. 1263–1277 (2015)
56. Welinder, P., Perona, P.: Online crowdsourcing: rating annotators and obtaining cost-effective labels. In: CVPR Workshop (ACVHL), pp. 25–32. IEEE (2010)
57. Whang, S.E., Lofgren, P., Garcia-Molina, H.: Question selection for crowd entity resolution. PVLDB **6**(6), 349–360 (2013)
58. Whitehill, J., Ruvolo, P., Wu, T., Bergsma, J., Movellan, J.R.: Whose vote should count more: Optimal integration of labels from labelers of unknown expertise. In: NIPS, pp. 2035–2043 (2009)
59. Yan, T., Kumar, V., Ganesan, D.: Crowdsearch: exploiting crowds for accurate real-time image search on mobile phones. In: MobiSys, pp. 77–90 (2010)
60. Ye, P., EDU, U., Doermann, D.: Combining preference and absolute judgements in a crowd-sourced setting. In: ICML Workshop (2013)
61. Zhang, C.J., Chen, L., Jagadish, H.V., Cao, C.C.: Reducing uncertainty of schema matching via crowdsourcing. PVLDB **6**(9), 757–768 (2013)
62. Zhang, C.J., Tong, Y., Chen, L.: Where to: Crowd-aided path selection. PVLDB **7**(14), 2005–2016 (2014)
63. Zheng, Y., Wang, J., Li, G., Cheng, R., Feng, J.: QASCA: A quality-aware task assignment system for crowdsourcing applications. In: SIGMOD, pp. 1031–1046 (2015)
64. Zhuo, H.H.: Crowdsourced action-model acquisition for planning. In: AAAI, pp. 3439–3446

Chapter 2
Crowdsourcing Background

This chapter introduces the background of crowdsourcing. Section 2.1 gives an overview of crowdsourcing, and Sect. 2.2 introduces the crowdsourcing workflow. Next, Sect. 2.3 introduces some widely used crowdsourcing platforms, and Sect. 2.4 discusses existing tutorials, surveys, and books on crowdsourcing. Finally, Sect. 2.5 presents the optimization goals of crowdsourced data management.

2.1 Crowdsourcing Overview

Crowdsourcing has several variants with different task granularities and worker incentive strategies. Figure 2.1 illustrates the crowdsourcing space with different characteristics.

Task Granularity Tasks can be classified into micro-tasks (e.g., labeling an image) and macro-tasks (e.g., translating a paper). A micro-task usually takes several seconds or minutes to finish, while a macro-task may take several hours or days. As the majority of existing crowdsourced data management works focus on micro-tasks, we restrict the scope of this book to micro-tasks as well. But we indeed believe that the studies of macro-tasks should be an important research topic in the future.

Worker Incentive Strategies Worker incentive strategies include money, entertainment, hidden, and volunteer. (1) The requester needs to pay workers who answer her tasks. For example, ImageNet uses crowdsourcing to build an image labeling dataset. It first constructs a concept hierarchy, e.g., animal, dog, and Husky, and then utilizes the crowd to link each image to a concept on the hierarchy. (2) The workers are interested in answering the tasks, e.g., games. For example, ESP game uses the game players to identify objects from images. As an another example, Foldit designs a game and utilizes the game users to identify the structure of proteins. (3) The tasks are hidden and workers do not know that they are answering requesters' tasks.

© Springer Nature Singapore Pte Ltd. 2018

G. Li et al., *Crowdsourced Data Management*,

https://doi.org/10.1007/978-981-10-7847-7_2

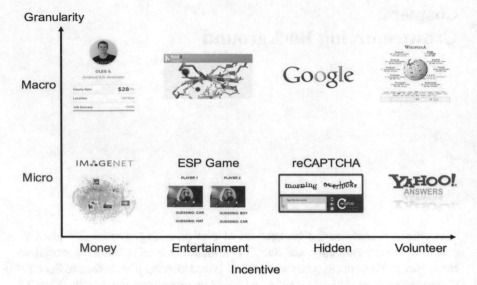

Fig. 2.1 Crowdsourcing space

For example, reCAPTCHA uses crowdsourcing to digitalize 20 million newspapers of *The New York Times*. As another example, Google utilizes users' click-through data to improve the ranking of search engines. (4) Workers volunteer to contribute toward the tasks. For example, in Yahoo! QA, users volunteer to answer questions. In Wikipedia, users volunteer to contribute articles.

This book focuses on micro-tasks where workers are paid to answer the tasks, e.g., on AMT and CrowdFlower. Micro-tasks and tasks will be interchangeably used in the remainder of this paper.

2.2   Crowdsourcing Workflow

In a crowdsourcing platform (e.g., AMT [1]), requesters submit tasks to the platform, which publishes the tasks; then, workers find interesting tasks, perform the selected tasks, and return the results. In the following, the life cycle of tasks from the individual perspective of requesters and workers will be discussed.

2.2.1   Workflow from Requester Side

Suppose a requester has an entity resolution problem to solve, which aims to find the same entity from 1,000 products. (1) The requester first designs the task by

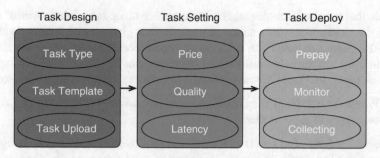

Fig. 2.2 Task design

using templates provided by the crowdsourcing platforms, calling APIs defined in crowdsourcing platforms, or writing codes to interact with crowdsourcing platforms. For example, if the requester uses the template, she can select the user interface of a task (e.g., showing a pair of products and asking the crowd to choose between "the same" and "different"). (2) The requester sets some properties of the tasks, e.g., the price of a task, the number of workers to answer a task, and the time duration to answer a task. (3) The requester publishes the tasks to the platform and collects the answers from the platform. Next the details of each step will be discussed (Fig. 2.2).

2.2.1.1 Task Design

There are several important task types that are widely used in real-world crowd-sourcing platforms.

Single Choice The task asks workers to select a single answer from multiple options. For example, in entity resolution, given a pair of objects, it asks the worker to select whether or not they refer to the same entity (options: Yes or No). In sentiment analysis, given a review, it asks the worker to assess the sentiment of the review (options: Positive, Neutral, Negative).

Multiple Choices The task asks workers to select multiple answers from multiple options. For example, given a picture, workers are asked to select the objects from a list of given objects (e.g., Monkey, Tree, Banana) that appear in the picture.

Fill-in-the-Blank. The task asks workers to fill in the values of some specified attributes. For example, given a professor, the task asks workers to provide the email and homepage of the professor. Given a mountain, the task asks workers to provide the height of the mountain. Thus the values provided by the workers are open, which can be numerical, textual, or categorical.

Collection The task asks workers to collect some information. For example, a task asks workers to provide NBA players and top 100 universities in the USA.

Note that the single-choice and multiple-choice tasks use a "close-world" model where the workers can only select answers from a given set of options, while the fill-in-the-blank and collection tasks use an "open-world" model where workers can provide open answers. Note that most operators can be implemented using these four task types, and there are multiple user interfaces to support each operator.

Real-World Tasks Next we introduce some real-world applications that can use these task types to support them.

Sentiment Analysis The aim is to identify the sentiment of a review, an image, or a conversation. The single-choice task type can be used, where a task asks workers to select an answer from positive, negative, and neutral.

Search Relevance The aim is to label the relevance of search results so as to rank the search results. The single-choice task type can be used, where each task asks workers to select an answer (or a rating) from highly relevant, relevant, neutral, irrelevant, and highly irrelevant (or 1, 2, 3, 4, 5).

Content Moderation The aim is to moderate the content of an image or a website. The single-choice task type can be used, where a task provides workers with an image and asks workers to check whether the image contains adult content, i.e., selecting an answer from Yes or No.

Data Collection The aim is to collect some unknown data. For example, many companies aim to get new points of interest (POIs). The collection task type can be used, where each task asks workers to provide new POIs.

Data Categorization The aim is to categorize an object to predefined concepts. Since an object can belong to multiple concepts, the multiple-choice task type can be used, where each task asks workers to select multiple answers from a given set of concepts (e.g., singer, actor, director).

Image/Audio Transcription The aim is to turn images and audio into useful data. The fill-in-the-blank task type can be used, where each task asks workers to provide the answers of a given image or audio.

After selecting the task type, requesters can select the user interface or write HTML code to design the tasks. Next, the requesters upload the data of the tasks and use the UI to render the tasks.

2.2.1.2 Task Settings

The requester also needs to set up some task settings based on her requirements. There are three main factors that the requester needs to consider.

Pricing The requester needs to set the price of each task. Usually task prices vary from a few cents to several dollars. Note that pricing is a complex game-theoretic problem. Usually, high prices can attract more workers, thereby reducing the latency; but paying more does not always improve answer quality [13]. Thus it is an interesting problem to set an appropriate price for a task.

Timing The requester can set time constraints for each task. For each task, the requester can set the time bound (e.g., 60 s) to answer it, and the worker must answer it within this time bound. The requester can also set the expiration time of the tasks, i.e., the maximum time that the tasks will be available to workers on the platform (e.g., 24 h).

Quality Control The requester can select the quality-control techniques provided by the crowdsourcing platform or design her own quality-control methods. For example, requesters can ask the workers to answer qualification tests by providing workers with some golden tasks with known answers, and the workers who cannot pass the qualification tests (i.e., having low quality on these golden tasks) cannot answer the normal tasks. Requesters can also utilize hidden tests that mix some golden tasks into the normal tasks (but the workers do not know there are some golden tasks), and workers that have low quality on golden tasks will be eliminated. The quality-control techniques will be discussed in Chap. 3.

2.2.1.3 Task Deployment

Requesters are required to prepay money to the crowdsourcing platform, where the monetary cost includes the cost for the workers and the cost for the platform. The worker cost can be computed by the number of tasks times the price of each task times the number of workers assigned for each task, and the platform cost is the worker cost times a platform ratio (e.g., 5%).

After prepaying, the requesters can publish the tasks to the platform. Then requesters can monitor the tasks, e.g., checking the percentage of completed tasks and checking the quality of returned answers. Finally, the requester can collect all the answers from the crowd. The requester can also evaluate the performance of workers who answered her tasks. The crowdsourcing platform assigns workers with a reputation based on these evaluations.

2.2.2 Workflow from Worker Side

From the workers' perspective, workers can browse and select the available tasks published by the requesters. When accepting a task, they have to finish the task within the time duration specified by the requesters. If a worker has accomplished a task, the requester who publishes the task can approve or disapprove the worker's answers. The approved workers will get paid from the requester. Workers can also be eliminated by the requester or platform, if they have low quality.

2.2.3   Workflow from Platform Side

When a requester deploys some tasks, the platform publishes the tasks on the platform, and workers can see the tasks on the platform. When a worker submits an answer, the platform collects the answers and sends them to the corresponding requesters. When a worker requests a task, the platform assigns a task to the worker. Usually the task is randomly assigned by default. In some crowdsourcing platform, requesters are allowed to control the task assignment, and the details will be introduced in Chap. 3.

2.3   Crowdsourcing Platforms

There are many open crowdsourcing platforms from which one can access crowd resources very early.

2.3.1   Amazon Mechanical Turk (AMT)

AMT [1] is a widely used crowdsourcing platform. AMT focuses on micro-tasks, e.g., labeling an image. A requester can group multiple tasks as a Human Intelligence Task (called HIT). The requester can also set some requirements, e.g., the price of a HIT, the time constraint for answering a HIT, the expiration time for a job to be available on AMT, and the qualification test.

A requester can build HITs from several different ways. (1) The requester can utilize user interface. AMT provides many templates, such as categorization, data collection, sentiment analysis, and image tagging. After designing the worker interface, the requester needs to upload the task files. (2) AMT Command Line Tools (CLT). AMT predefines a set of commands in CLT, such as loadHITs, getResults, grantBonus, blockWorker, and so on. The requester can utilize CLT to easily build HITs by specifying three files: the data file, the task file to specify user interface for the tasks, and the property file to add the title, description, keywords, reward, and assignments for the HIT. The CLT is suitable when a requester has a relatively small number of assignments. (3) AMT APIs. There are many APIs in AMT, including the creation of HITs, blocking/unblocking the workers, collection of the finished answers, and statistics collection for the requester and workers. There are three steps to use the APIs to publish tasks: (i) download the SDK for a specified language, e.g., python and java; (ii) specify the title, the description, the reward, the content of tasks, and the detailed properties for the HIT; and (iii) publish the HITs to the platform. (4) The requester can build her own server to manage the tasks and embed the tasks into AMT using innerHTML. When a worker requires a task, AMT transforms the request to the requester's server, and then the requester can decide how to assign tasks to the server. When a worker submits an answer to AMT, AMT also transforms the result to the requester. Thus the requesters can control the task assignment using the fourth strategy in AMT.

A worker can browse HITs on AMT. Each HIT has some information, e.g., the description of the task, the price, the keywords, the qualification test if required, and the requester's id. After a worker submits the answers of HITs to the platform, she can find the total earnings and the status of the submitted HITs on the platform.

2.3.2 CrowdFlower

CrowdFlower [3] has similar functionalities with AMT, but they still have some differences. First, CrowdFlower has a quality-control component, and it leverages the hidden test (injecting some golden tasks) (see Chap. 3) to block low-quality workers. Second, besides publishing the tasks on its own platform, CrowdFlower also publishes the tasks to other platforms.

CrowdFlower has two ways to allow requesters to design tasks. (1) Requesters can utilize the APIs to inspect the properties for existing tasks and create new tasks. Currently, there are Python and Ruby packages for the usage of APIs. (2) Requesters can use a user interface with six steps. The first step is to upload data. In this step, the requester should firstly specify the format of the data set and then upload them with a proper file format. The second step is to design tasks. Requesters can design the layout of the task in the dashboard or directly write CML (CrowdFlower Markup Language) code in the code-writing box. The third step is the creation of the test questions. Test questions are questions with known answers. In the fourth step, requesters can choose to publish their tasks. At this step, the requester should specify the number of tasks to launch and set the payment amount per page. The last two steps are task monitor to the answer statistic and answer collection.

Workers in CrowdFlower can directly find tasks on the dashboard in the worker user interface. The dashboard presents the details about the unfinished tasks, which include task title, requirements for worker skills, reward, the number of remaining tasks, and the workers satisfaction for the task. And there is also a flag that indicates whether the task includes qualification test. If the task has qualification-test questions, any worker who wants to participate in this task has to pass this qualification test first. CrowdFlower also allows requesters to utilize the hidden test (mixing some golden tasks into the normal tasks) to reject workers with low quality on golden tasks.

2.3.3 Other Platforms

There are other crowdsourcing platforms. ChinaCrowd [2] is a multilingual crowd-sourcing platform, which supports Chinese and English. It also supports spatial

Table 2.1 Crowdsourcing platforms

	Task type	Worker	Requester		
		Task search	Qualification	Hidden	Task assignment by requesters
AMT	Micro	✓	✓	✓	✓
CrowdFlower	Micro	✓	✓	✓	×
ChinaCrowd	Micro	✓	✓	✓	✓
Upwork	Macro	✓	×	×	×
Crowdspring	Macro	✓	×	×	×
Zhubajie	Macro	✓	×	×	×

crowdsourcing. Upwork [5], Zhubajie [6], and Crowdspring [4] can support macro-tasks, e.g., translating a paper or developing a mobile application. gMission [9] is a spatial crowdsourcing platform that supports spatial tasks.

Table 2.1 shows the differences among different crowdsourcing platforms.

2.4 Existing Surveys, Tutorials, and Books

There are some existing crowdsourcing tutorials (e.g., in VLDB'16 [7], VLDB'15 [14], ICDE'15 [8], VLDB'12 [10]), and most of them focus on a specific part of crowdsourcing. VLDB'16 [7] investigates human factors involved in task assignment and completion. VLDB'15 [14] focuses on truth inference in quality control. ICDE'15 [8] reviews individual crowdsourcing operators, crowdsourced data mining, and social applications. VLDB'12 [10] introduces crowdsourcing platforms and discusses design principles for crowdsourced data management. Doan et al. [11] give a survey on general crowdsourcing. Adam Marcus and Aditya G. Parameswaran [20] discuss crowdsourced data management mainly from the industry view. Compared with these tutorials and surveys, we will summarize an overview of a wide spectrum of work on crowdsourced data management, with a special focus on the fundamental techniques for controlling quality, cost, and latency. We will also introduce crowdsourcing systems and operators, including the works published very recently [12, 15–19, 21–29], which can provide a detailed summary on crowdsourcing data management techniques.

2.5 Optimization Goal of Crowdsourced Data Management

There are several optimization goals in crowdsourced data management.

Improving The Quality The crowd may return noisy results. The goal is to tolerate the crowd errors and improve the quality of crowdsourcing. The quality-control methods will be discussed in Chap. 3.

Reducing The Cost The crowd is not free. The goal is to reduce the crowd cost. The details on cost control will be introduced in Chap. 4.

Decreasing The Latency Workers may involve high latency due to several reasons: (1) workers are not interested in the tasks; (2) workers are not available in the current time. The goal is to decrease the latency. The details of latency control will be discussed in Chap. 5.

Optimizing Crowdsourced Database Operators and Systems. Crowdsourcing platforms are hard to use. The goal is to design effective crowdsourcing systems to facilitate requesters to easily deploy their applications. In addition, optimization techniques should be designed to optimize the three goals in crowdsourcing. We discuss the details on crowdsourcing database systems and optimization in Chap. 6. Crowdsourced operators are also required to support various applications, and the details will be introduced in Chap. 7.

References

1. Amazon mechanical turk. https://www.mturk.com/
2. Chinacrowd. http://www.chinacrowds.com
3. Crowdflower. http://www.crowdflower.com
4. Crowdspring. https://www.crowdspring.com
5. Upwork. https://www.upwork.com
6. Zhubajie. http://www.zbj.com
7. Amer-Yahia, S., Roy, S.B.: Human factors in crowdsourcing. PVLDB **9**(13), 1615–1618 (2016)
8. Chen, L., Lee, D., Milo, T.: Data-driven crowdsourcing: Management, mining, and applications. In: ICDE, pp. 1527–1529. IEEE (2015)
9. Chen, Z., Fu, R., Zhao, Z., Liu, Z., Xia, L., Chen, L., Cheng, P., Cao, C.C., Tong, Y., Zhang, C.J.: gmission: a general spatial crowdsourcing platform. PVLDB **7**(13), 1629–1632 (2014)
10. Doan, A., Franklin, M.J., Kossmann, D., Kraska, T.: Crowdsourcing applications and platforms: A data management perspective. PVLDB **4**(12), 1508–1509 (2011)
11. Doan, A., Ramakrishnan, R., Halevy, A.Y.: Crowdsourcing systems on the world-wide web. Commun. ACM **54**(4), 86–96 (2011)
12. Fan, J., Li, G., Ooi, B.C., Tan, K., Feng, J.: icrowd: An adaptive crowdsourcing framework. In: SIGMOD, pp. 1015–1030 (2015)
13. Faradani, S., Hartmann, B., Ipeirotis, P.G.: What's the right price? pricing tasks for finishing on time. In: AAAI Workshop (2011)
14. Gao, J., Li, Q., Zhao, B., Fan, W., Han, J.: Truth discovery and crowdsourcing aggregation: A unified perspective. PVLDB **8**(12), 2048–2049 (2015)
15. Groz, B., Milo, T.: Skyline queries with noisy comparisons. In: PODS, pp. 185–198 (2015)
16. Haas, D., Ansel, J., Gu, L., Marcus, A.: Argonaut: Macrotask crowdsourcing for complex data processing. PVLDB **8**(12), 1642–1653 (2015)
17. Haas, D., Wang, J., Wu, E., Franklin, M.J.: Clamshell: Speeding up crowds for low-latency data labeling. PVLDB **9**(4), 372–383 (2015)
18. Joglekar, M., Garcia-Molina, H., Parameswaran, A.G.: Comprehensive and reliable crowd assessment algorithms. In: ICDE, pp. 195–206 (2015)
19. Ma, F., Li, Y., Li, Q., Qiu, M., Gao, J., Zhi, S., Su, L., Zhao, B., Ji, H., Han, J.: Faitcrowd: Fine grained truth discovery for crowdsourced data aggregation. In: KDD, pp. 745–754 (2015)

20. Marcus, A., Parameswaran, A.G.: Crowdsourced data management: Industry and academic perspectives. Foundations and Trends in Databases **6**(1–2), 1–161 (2015)
21. Ouyang, W.R., Kaplan, L.M., Martin, P., Toniolo, A., Srivastava, M.B., Norman, T.J.: Debiasing crowdsourced quantitative characteristics in local businesses and services. In: IPSN, pp. 190–201 (2015)
22. Verroios, V., Garcia-Molina, H.: Entity resolution with crowd errors. In: ICDE, pp. 219–230 (2015)
23. Verroios, V., Lofgren, P., Garcia-Molina, H.: tdp: An optimal-latency budget allocation strategy for crowdsourced MAXIMUM operations. In: SIGMOD, pp. 1047–1062 (2015)
24. Wang, S., Xiao, X., Lee, C.: Crowd-based deduplication: An adaptive approach. In: SIGMOD, pp. 1263–1277 (2015)
25. Zhao, Z., Wei, F., Zhou, M., Chen, W., Ng, W.: Crowd-selection query processing in crowdsourcing databases: A task-driven approach. In: EDBT, pp. 397–408 (2015)
26. Zheng, Y., Cheng, R., Maniu, S., Mo, L.: On optimality of jury selection in crowdsourcing. In: EDBT, pp. 193–204 (2015)
27. Zheng, Y., Li, G., Cheng, R.: DOCS: domain-aware crowdsourcing system. PVLDB **10**(4), 361–372 (2016)
28. Zheng, Y., Li, G., Li, Y., Shan, C., Cheng, R.: Truth inference in crowdsourcing: Is the problem solved? PVLDB **10**(5), 541–552 (2017)
29. Zheng, Y., Wang, J., Li, G., Cheng, R., Feng, J.: QASCA: A quality-aware task assignment system for crowdsourcing applications. In: SIGMOD, pp. 1031–1046 (2015)

Chapter 3
Quality Control

The results collected from crowd workers may not be reliable because (1) there are some malicious workers that randomly return the answers and (2) some tasks are hard and workers may not be good at these tasks. Thus it is important to exploit the different characteristics of workers and tasks and control the quality in crowdsourcing. Existing studies propose various quality-control techniques to address these issues.

In this chapter, an overview of quality control is given in Sect. 3.1; then the techniques on truth inference and task assignment are discussed in Sects. 3.2 and 3.3, respectively. Finally, the chapter is summarized in Sect. 3.4.

3.1 Overview of Quality Control

To give an overview of crowdsourced quality, we show the crowdsourcing framework in Fig. 3.1. The requester first deploys tasks and identifies her budget to the crowdsourcing platform (e.g., AMT). To control the quality, each task is usually assigned to multiple workers and the answer of the task is aggregated by the answers from these assigned workers. Next the workers will come to interact with the crowdsourcing platform in two components:

(1) **Truth Inference,** which collects workers' answers and infers the truth of each task by leveraging all collected answers from workers. Note that workers may yield low quality or even noisy answers, e.g., a malicious worker will intentionally give wrong answers; workers may also have different levels of expertise, and an untrained worker may be incapable of accomplishing certain tasks. Thus, in order to achieve high quality, we need to tolerate crowd errors and infer high-quality results from noisy answers.

(2) **Task Assignment,** which assigns tasks to workers. Since workers may have different backgrounds and diverse qualities on tasks, an intelligent task assignment

© Springer Nature Singapore Pte Ltd. 2018
G. Li et al., *Crowdsourced Data Management*,
https://doi.org/10.1007/978-981-10-7847-7_3

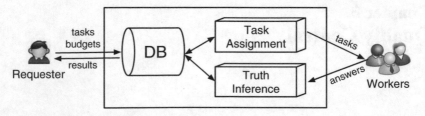

Fig. 3.1 Crowdsourcing quality-control framework

algorithm will judiciously select suitable tasks to appropriate workers. Existing works study the task assignment problem by focusing on two settings, based on the perspectives of tasks and workers:

- *Task Assignment Setting.* In this setting, when a worker comes, the focus is on studying which subset of tasks should be assigned to the coming worker. This problem is often called the *online task assignment problem* [7, 17, 18, 52]. Existing platforms such as AMT support assigning tasks in the task assignment setting. To be specific, the requester builds a server to control the tasks. Then the requester could use the "external HIT" [4] way provided by AMT, which embeds the generated HTML pages by the requester's server into AMT's web frame, and workers directly interact with the server through the web frame. Next when a worker comes, the server can identify the worker from the individual worker-id (a string of 14 characters) provided by AMT. Finally, the requester can dynamically batch the selected tasks in a HIT and assign the task to the coming worker.
- *Worker Selection Setting.* In this setting, given a task and a set of candidate workers, the focus is on studying which subset of workers should be selected to answer the task. The problem is often called the *jury selection problem* [9, 49]. With the increasing amount of data, large-scale Internet companies such as Google and Facebook require human experts to label the data, e.g., they want to be sure whether there is a Starbucks coffee in a specific point of interest (POI) location. Workers with different qualities and budget requirements come to answer the task, and the problem aims at wisely selecting a subset of workers, such that the overall quality is maximized within a given budget constraint.

Most of the existing quality control techniques focus on truth inference and task assignment. Next, we introduce the techniques of truth inference (Sect. 3.2) and task assignment (Sect. 3.3), respectively.

3.2 Truth Inference

Section 3.2.1 first defines the truth inference problem, then Sect. 3.2.2 illustrates a unified solution framework adopted by existing works, and next Sect. 3.2.3 gives the differences of different methods. Finally, several possible extensions of truth inference are introduced in Sect. 3.2.4.

3.2.1 Truth Inference Problem

Definition 3.1 (Task) A task set $\mathcal{T}$ contains n tasks, i.e., $\mathcal{T} = \{t_1, t_2, \ldots, t_n\}$. Each task asks workers to provide an answer.

Existing studies mainly focus on three types of tasks.

Decision-Making Tasks A decision-making task has a claim and asks workers to make a decision on whether the claim is *true* (denoted as "T") or *false* (denoted as "F"). For example, the entity resolution task set $\mathcal{T}$ in Table 3.2 contains six decision-making tasks. Decision-making tasks are widely used and studied in existing crowdsourcing works [5, 10, 11, 13, 22, 23, 26, 29–31, 37, 40, 43, 44, 53] because of its conceptual simplicity and widespread applications.

Next we take entity resolution as an example, which tries to find pairs of products in Table 3.1 that refer to the same real-world entity. A straightforward way is to generate a task set $\mathcal{T} = \{(r_1 = r_2), (r_1 = r_3), (r_1 = r_4), (r_2 = r_3), (r_2 = r_4), (r_3 = r_4)\}$ with $n = 6$ decision-making tasks, where each task has two choices: (*true, false*), and asks workers to select a choice for the task. For example, t_2 (or $r_1 = r_3$) asks whether the claim "*iPad Two 16GB WiFi White = Apple iPhone 4 16GB White*" is *true* (T) or *false* (F). Tasks are then published to crowdsourcing platforms (e.g., AMT [1]) and workers' answers are collected.

Table 3.1 An example product dataset (for crowdsourced entity resolution)

ID	Product name
r_1	*iPad Two 16GB WiFi White*
r_2	*iPad 2nd generation 16GB WiFi White*
r_3	*Apple iPhone 4 16GB White*
r_4	*iPhone 4th generation White 16GB*

Table 3.2 The collected workers' answers for all tasks (i.e., asking whether two products in Table 3.1 refer to the same entity)

	t_1: $(r_1=r_2)$	t_2: $(r_1=r_3)$	t_3: $(r_1=r_4)$	t_4: $(r_2=r_3)$	t_5: $(r_2=r_4)$	t_6: $(r_3=r_4)$
w_1	F	T	T	F	F	F
w_2		F	F	T	T	F
w_3	T	F	F	F	F	T

Single-Choice (and Multiple-Choice) Tasks A single-choice task contains a question and a set of candidate choices and asks workers to select a single choice out of the candidate choices. For example, in sentiment analysis, a task asks workers to select the sentiment (*positive, neutral, negative*) of a given tweet. A decision-making task is a special case of a single-choice task, with two special choices ("T" and "F"). Single-choice tasks are especially studied in [5, 10, 11, 23, 26, 30, 31, 37, 40, 44, 53]. A direct extension of single-choice tasks is multiple-choice tasks, where workers can select multiple choices (not only a single choice) out of a set of candidate choices. For example, in image tagging, given a set of candidate tags for an image, workers are asked to select the tags that the image contains. However, as addressed in [34, 52], a multiple-choice task can be easily transformed to a set of decision-making tasks, e.g., for an image tagging task (multiple-choice), each transformed decision-making task asks whether or not a tag is contained in an image. Thus the methods in decision-making tasks can be directly extended to handle multiple-choice tasks.

Numeric Tasks Numeric tasks ask workers to provide a value for a question. For example, a task could ask about the height of Mount Everest. Different from the tasks above, workers' inputs are "open" and workers can provide any values (but in the above methods, workers must select answers from some given choices). Most of these tasks focus on numeric values, which have inherent orderings (e.g., compared with 8800 m, 8845 m is closer to 8848 m). Existing works [26, 37] especially study such tasks by considering the inherent orderings between values.

Others Besides the above tasks, there are other types of tasks, e.g., translate a language to another [8] or ask workers to collect data (e.g., the name of a celebrity) [16, 41]. However, it is hard to control the quality for such "open" tasks. Thus they are rarely studied in existing works [8, 16, 41]. In this section, we focus only on the above three tasks, and readers are referred to the papers [8, 16, 41] for more details on other tasks.

Definition 3.2 (Worker) Let $\mathcal{W} = \{w\}$ denote a set of workers, $\mathcal{W}^i$ denote the set of workers that have answered task t_i, and $\mathcal{T}^w$ denote the set of tasks that have been answered by a worker w.

Definition 3.3 (Answer) Each task t_i can be answered with a subset of workers in $\mathcal{W}$, and let v_i^w denote the worker w's answer for task t_i. Let $V = \{v_i^w\}$ denote the set of answers collected from all workers for all tasks.

Table 3.2 shows an example, with answers to $\mathcal{T}$ given by three workers $\mathcal{W} = \{w_1, w_2, w_3\}$. (The empty cell means that the worker does not answer the task.) For example, $v_4^{w_1} = $ F means worker w_1 answers t_4 (i.e., $r_2 = r_3$) with "F", i.e., w_1 thinks that $r_2 \neq r_3$. The set of workers that answer t_1 is $\mathcal{W}^1 = \{w_1, w_3\}$, and the set of tasks answered by worker w_2 is $\mathcal{T}^{w_2} = \{t_2, t_3, t_4, t_5, t_6\}$.

Definition 3.4 (Truth) Each task t_i has a true answer, called the *ground truth* (or *truth*), denoted as v_i^*.

Table 3.3 Notations used in Sect. 3.2

Notation	Description
t_i	The i-th task ($1 \leq i \leq n$) and $\mathcal{T} = \{t_1, t_2, \ldots, t_n\}$
w	The worker w and $\mathcal{W} = \{w\}$ is the set of workers
$\mathcal{W}^i$	The set of workers that have answered task t_i
$\mathcal{T}^w$	The set of tasks that have been answered by worker w
v_i^w	The answer given by worker w for task t_i
V	The set of workers' answers for all tasks, i.e., $V = \{v_i^w\}$
v_i^*	The (ground) truth for task t_i ($1 \leq i \leq n$)

For the example task set $\mathcal{T}$ in Table 3.1, only pairs ($r_1 = r_2$) and ($r_3 = r_4$) are *true*, and thus $v_1^* = v_6^* = \mathrm{T}$, and others' truth are F.

Based on the above notations, the truth inference problem is to infer the (unknown) truth v_i^* for each task t_i based on V.

Definition 3.5 (Truth Inference in Crowdsourcing) Given workers' answers V, infer the truth v_i^* of each task $t_i \in \mathcal{T}$.

Table 3.3 summarizes the notations used in this section.

3.2.2 Unified Solution Framework

A naive solution is majority voting (MV) [16, 33, 35], which regards the choice answered by the majority of workers as the truth. Based on Table 3.2, the truth derived by MV is $v_i^* = \mathrm{F}$ for $2 \leq i \leq 6$, and especially it randomly infers v_1^* to break the tie. The MV incorrectly infers v_6^*, and has 50% chance to infer v_1^* wrongly. The reason is that MV assumes that each worker has the same quality, and in reality, workers have different qualities: some are experts or ordinary workers, while others are spammers (who randomly answer tasks in order to receive money) or even malicious workers (who intentionally give wrong answers). Taking a closer look at Table 3.2, it can be observed that w_3 has a higher quality, and the reason is that if t_1 is not considered (which receives one "T" and one "F"), then w_3 gives four out of five answers that are reported by the majority of workers, while w_1 and w_2 give both three out of five majority answers; thus it is reasonable to give higher trust to w_3's answer, and in this way all tasks' truth could be inferred correctly.

Based on the above discussions, existing works [5, 10, 11, 13, 22, 23, 26, 29–31, 37, 40, 43, 44, 53, 54] propose various ways to model a worker's quality. Although qualification tests and hidden tests can help to estimate a worker's quality, they require to label tasks with the truth beforehand, and a worker also requires to answer these "extra" tasks. To address this problem, existing works [5, 10, 11, 13, 22, 23, 26, 29–31, 37, 40, 43, 44, 53, 54] estimate each worker's quality purely based on workers' answers V. Intuitively, they capture the inherent relations between

---

Algorithm 1 Solution framework

Input: workers' answers V
Output: inferred truth v_i^* $(1 \leq i \leq n)$, worker quality q^w $(w \in \mathcal{W})$
 1: Initialize all workers' qualities (q^w for $w \in \mathcal{W}$);
 2: **while true do**
 3:      *// Step 1: Inferring the Truth*
 4:      **for** $1 \leq i \leq n$ **do**
 5:          Inferring the truth v_i^* based on V and $\{q^w \mid w \in \mathcal{W}\}$;
 6:      **end for**
 7:      *// Step 2: Estimating Worker Quality*
 8:      **for** $w \in \mathcal{W}$ **do**
 9:          Estimating the quality q^w based on V and $\{v_i^* \mid 1 \leq i \leq n\}$;
10:      **end for**
11:      *// Check for Convergence*
12:      **if** Converged **then**
13:          **break**;
14:      **end if**
15: **end while**
16: **return** v_i^* for $1 \leq i \leq n$ and q^w for $w \in \mathcal{W}$;

---

workers' qualities and tasks' truth: for a task, the answer given by a high-quality worker is highly likely to be the truth; conversely, for a worker, if the worker often correctly answers tasks, then the worker will be assigned with a high quality. By capturing such relations, they adopt an iterative approach, which jointly infers both the workers' qualities and tasks' truth.

By capturing the above relations, the general approach adopted by most of existing works [5, 10, 11, 13, 22, 23, 26, 29–31, 37, 40, 43, 44, 53, 54] is shown in Algorithm 1. The quality of each worker $w \in \mathcal{W}$ is denoted as q^w. In Algorithm 1, the method first initializes each worker's quality (line 1) and then adopts an iterative approach with two steps (lines 3–13). Next, we discuss the initialization, the two iterative steps, and the convergence, respectively.

Initialization (line 1): To initialize each worker's quality, there are different scenarios to consider:

- **If there is no known ground truth.**
  In this case, each worker's quality can be initialized randomly, or initialized as a perfect worker [20].
- **If a small set of tasks' ground truth are known.**
  In this case, we can leverage the golden tasks (Definition 3.6) to estimate each worker's quality. The basic idea is that we can estimate each worker's quality by comparing the worker's answers to the ground truth.

Definition 3.6 (Golden Tasks) Golden tasks are defined as tasks whose ground truths are known in advance.

In fact, there are two types of methods to utilize golden tasks: *qualification test* (Definition 3.7) and *hidden test* (Definition 3.8).

Definition 3.7 (Qualification Test) The *qualification test* is a form of test for the workers, where each worker is required to perform a set of golden tasks before she can actually answer tasks.

Definition 3.8 (Hidden Test) The *hidden test* is a form of test for the workers, where the golden tasks are mixed into the tasks and the workers do not know which are golden tasks.

Finally, a worker's quality is computed based on her answering performance on these golden tasks. Then based on the worker's answering performance on *golden tasks*, the worker's quality can be computed.

Step 1: **Inferring the Truth** (lines 3–6): infer each task's truth based on workers' answers and qualities. In this step, different task types are handled differently. Furthermore, some existing works [43, 44] explicitly model each task, e.g., [44] which regards that different tasks may have different difficulties. We discuss how existing works model a task in Sect. 3.2.3.1.

Step 2: **Estimating Worker Quality** (lines 7–10): based on workers' answers and each task's truth (derived from Step 1), estimate each worker's quality. In this step, existing works model each worker w's quality q^w differently. For example, [5, 11, 22, 29] model q^w as a single value, while [10, 23, 29, 37, 40] model q^w as a matrix. We discuss worker's models in Sect. 3.2.3.2.

Convergence (lines 11–14): the two iterations will run until convergence. Typically to identify convergence, existing works will check whether the change of two sets of parameters (i.e., workers' qualities and tasks' truth) is below some defined threshold (e.g., 10^{-3}). Finally the inferred truth and workers' qualities are returned.

Running Example Let us show how the method **PM** [5, 27] works for Table 3.2. **PM** models each worker w as a single value $q^w \in [0, +\infty)$, and a higher value implies a higher quality. Initially, each worker $w \in \mathcal{W}$ is assigned with the same quality $q^w = 1$. Then the two steps devised in **PM** are as follows:

Step 1 (line 5): $v_i^* = \text{argmax}_v \sum_{w \in \mathcal{W}^i} q^w \cdot \mathbb{1}_{\{v = v_i^w\}};$

Step 2 (line 9): $q^w = -\log\left(\dfrac{\sum_{t_i \in \mathcal{T}^w} \mathbb{1}_{\{v_i^* \neq v_i^w\}}}{\max_{w \in \mathcal{W}}\{\sum_{t_i \in \mathcal{T}^w} \mathbb{1}_{\{v_i^* \neq v_i^w\}}\}}\right).$

The indicator function $\mathbb{1}_{\{\cdot\}}$ returns 1 if the statement is true; 0, otherwise. For example, $\mathbb{1}_{\{5=3\}} = 0$ and $\mathbb{1}_{\{5=5\}} = 1$. For the 1st iteration, in step 1, compute each task's truth from workers' answers by considering which choice receives the highest aggregated workers' qualities. Intuitively, the answer given by many high-quality workers is likely to be the truth. For example, for task t_2, as it receives one T and two F's from workers and each worker is of the same quality, then $v_2^* = F$. Similarly we get $v_1^* = T$ and $v_i^* = F$ for $2 \leq i \leq 6$. In step 2, based on the computed truth in step 1, give a high (low) quality to a worker if the worker makes few (a lot of) mistakes. For example, as the number of mistakes (i.e., $\sum_{t_i \in \mathcal{T}^w} \mathbb{1}_{\{v_i^* \neq v_i^w\}}$) for workers w_1, w_2, w_3 are 3, 2, 1, respectively, the computed qualities are $q^{w_1} = -\log(3/3) = 0$, $q^{w_2} = -\log(2/3) = 0.41$, and $q^{w_3} = -\log(1/3) = 1.10$. Following these two steps, the process will then iterate until convergence. In the converged results, the ground truths are $v_1^* = v_6^* = $ T, and $v_i^* = $ F $(2 \le i \le 5)$; the qualities are $q^{w_1} = 4.9 \times 10^{-15}$, $q^{w_2} = 0.29$, and $q^{w_3} = 16.09$. From the example, we can observe that PM can derive the truth correctly, and w_3 has a higher quality compared with w_1 and w_2.

3.2.3 Comparisons of Existing Works

Based on the unified framework, existing works [5, 10, 11, 13, 22, 23, 26, 29–31, 37, 40, 43, 44, 53, 54] can be categorized based on the following three factors:

Task Modeling: how existing works model a task (e.g., the difficulty of a task, the latent topics in a task).

Worker Modeling: how existing works model a worker's quality (e.g., worker probability, diverse skills).

Applied Techniques: how existing works incorporate task models and worker models in solving the truth inference problem.

We summarize how existing works [5, 10, 11, 13, 22, 23, 26, 29–31, 37, 40, 43, 44, 51, 53, 54] can be categorized based on the above factors in Table 3.4. Next we analyze each factor, i.e., task modeling, worker modeling, and applied techniques in Sects. 3.2.3.1, 3.2.3.2 and 3.2.3.3, respectively.

3.2.3.1 Task Modeling

- **Task Difficulty.** Different from most existing works which assume that a worker has the same quality for answering different tasks, some recent works [31, 44] model the difficulty in each task. They assume that each task has its difficulty level, and the more difficult a task is, the harder a worker can correctly answer the task. For example, in [44], the probability that worker w correctly answers task t_i is modeled as follows: $\Pr(v_i^w = v_i^* \mid d_i, q^w) = 1/(1 + e^{-d_i \cdot q^w})$, where $d_i \in (0, +\infty)$ represents the difficulty for task t_i, and the higher d_i is, the easier task t_i is. Intuitively, for a fixed worker quality $q^w > 0$, an easier task (high value of d_i) leads to a higher probability that the worker correctly answers the task.
- **Latent Topics.** Different from modeling each task as a value (e.g., difficulty), some recent works [13, 31, 43, 47, 50] model each task as a vector with K values. The basic idea is to exploit the diverse topics in a task, where the topic number (i.e., K) is predefined. For example, existing studies [13, 31] make use of the text description in each task and adopt topic model techniques [6, 46] to generate a vector of size K for the task [50], leverage entity linking techniques [39] (basically extract entities from each task and map them to knowledge bases), and infer a domain vector for each task, which captures the relatedness of the task to each specific domain, while Multi [43] learns a K-size vector without referring to external information (e.g., text descriptions).

Table 3.4 Comparisons of different truth inference methods

Method	Task types	Task modeling	Worker modeling	Applied techniques
MV	DM, SC	No model	No model	Straightforward
Mean	Numeric	No model	No model	Straightforward
Median	Numeric	No model	No model	Straightforward
ZC [11]	DM, SC	No model	Worker probability	PGM
GLAD [44]	DM, SC	Task difficulty	Worker probability	PGM
D&S [10]	DM, SC	No model	Confusion matrix	PGM
Minimax [53]	DM, SC	No model	Diverse skills	Optimization
BCC [23]	DM, SC	No model	Confusion matrix	PGM
CBCC [40]	DM, SC	No model	Confusion matrix	PGM
LFC [37]	DM, SC	No model	Confusion matrix	PGM
CATD [26]	DM, SC, Numeric	No Model	Worker probability confidence	Optimization
PM [5, 27]	DM, SC, Numeric	No model	Worker probability	Optimization
Multi [43]	DM	Latent topics	Diverse skills Worker bias Worker variance	PGM
KOS [22]	DM	No model	Worker probability	PGM
VI-BP [29]	DM	No model	Confusion matrix	PGM
VI-MF [29]	DM	No model	Confusion matrix	PGM
LFC_N [37]	Numeric	No model	Worker variance	PGM

Note: In the table, "DM" means "decision-making", "SC" means "single-choice", and "PGM" means "probabilistic graphical model"

3.2.3.2   Worker Modeling

- **Worker Probability** Worker probability uses a single real number (between 0 and 1) to model a worker w's quality $q^w \in [0, 1]$, which represents the ability that worker w correctly answers a task. The higher q^w is, the worker w has higher ability to correctly answer tasks. The model has been widely used in existing works [5, 11, 22, 29]. Some recent works [27, 44] extend the worker probability to model a worker's quality in a wider range, e.g., $q^w \in (-\infty, +\infty)$, and a higher q^w indicates the worker w's higher quality in answering tasks.
- **Confusion Matrix** Confusion matrix [10, 23, 29, 37, 40] is used to model a worker's quality for answering single-choice tasks. Suppose each task in $\mathcal{T}$ has ℓ fixed choices, then the confusion matrix q^w is an $\ell \times \ell$ matrix

$$
q^w = \begin{bmatrix}
q_{1,1}^w, \ q_{1,2}^w, \ \ldots, \ q_{1,\ell}^w \\
q_{2,1}^w, \ q_{2,2}^w, \ \ldots, \ q_{2,\ell}^w \\
\vdots \quad \vdots \quad \ddots \quad \vdots \\
q_{\ell,1}^w, \ q_{\ell,2}^w, \ \ldots, \ q_{\ell,\ell}^w
\end{bmatrix},
$$

where the j-th $(1 \leq j \leq \ell)$ row, i.e., $q_{j,\cdot}^w = [q_{j,1}^w, q_{j,2}^w, \ldots, q_{j,\ell}^w]$, represents the probability distribution of worker w's possible answers for a task if the truth of the task is the j-th choice. Each element $q_{j,k}^w$ $(1 \leq j \leq \ell, 1 \leq k \leq \ell)$ means that "given the truth of a task is the j-th choice, the probability that worker w selects the k-th choice," i.e., $q_{j,k}^w = \Pr(v_i^w = k \mid v_i^* = j)$ for any $t_i \in \mathcal{T}$. For example, decision-making tasks ask workers to select "T" (1st choice) or "F" (2nd choice) for each claim $(\ell = 2)$, then an example confusion matrix for w is
$$q^w = \begin{bmatrix} 0.8 & 0.2 \\ 0.3 & 0.7 \end{bmatrix},$$ where $q_{1,2}^w = 0.2$ means that if the truth of a task is "T," the probability that the worker answers "F" is 0.2.

- **Worker Bias and Worker Variance** Worker bias and variance [37, 43] are proposed to handle numeric tasks, where worker bias captures the effect that a worker may underestimate (or overestimate) the truth of a task, and worker variance captures the variation of errors around the bias. For example, given a set of photos with humans, each numeric task asks workers to estimate the height of the human on it. Suppose a worker w is modeled with bias τ_w and variance σ_w, then the answer v_i^w given by worker w is modeled to draw from the Gaussian distribution: $v_i^w \sim \mathcal{N}(v_i^* + \tau_w, \sigma_w)$, that is, (1) a worker with bias $\tau_w \gg 0$ ($\tau_w \ll 0$) will overestimate (underestimate) the height, while $\tau_w \to 0$ leads to more accurate estimation; (2) a worker with variance $\sigma_w \gg 0$ means a large variation of error, while $\sigma_w \to 0$ leads to a small variation of error.

- **Confidence** Existing works [21, 26] observe that if a worker answers plenty of tasks, then the estimated quality for the worker is confident; otherwise, if a worker answers only a few tasks, then the estimated quality is not confident. Inspired by this observation, [31] assigns higher qualities to the workers who answer plenty of tasks than the workers who answer few tasks. To be specific, for a worker w, the method uses the chi-square distribution [2] with 95% confidence interval, i.e., $\mathcal{X}_{(0.975,|\mathcal{T}^w|)}^2$ as a coefficient to scale up the worker's quality, where $|\mathcal{T}^w|$ is the number of tasks that worker w has answered. $\mathcal{X}_{(0.975,|\mathcal{T}^w|)}^2$ increases with $|\mathcal{T}^w|$, i.e., the more tasks w has answered, the higher the worker w's quality is scaled to.

- **Diverse Skills** A worker may have various levels of expertise for different topics. For example, a sports fan that rarely pays attention to entertainment may answer tasks related to *sports* more correctly than tasks related to *entertainment*. Different from most of the above models which have an assumption that a worker has the *same* quality to answer different tasks, existing works [13, 31, 43, 47, 50, 53] model the diverse skills in a worker and capture a worker's diverse qualities for different tasks. The basic ideas of [13, 53] are that they model a worker w's quality as a vector of size n, i.e., $q^w = [q_1^w, q_2^w, \ldots, q_n^w]$, where q_i^w indicates worker w's quality for task t_i. Different from [13, 53], some recent works [31, 43, 47, 50] model a worker's quality for different latent topics, i.e., $q^w = [q_1^w, q_2^w, \ldots, q_K^w]$, where the number K is predefined, indicating the number of latent topics. They [31, 43, 47, 50] assume that each task is related to one or more topics in these K latent topics, and a worker is highly probable to correctly answer a task if the worker has a high quality in the task's related topics.

3.2.3.3   Applied Techniques

Existing works [5, 10, 11, 13, 22, 23, 26, 29–31, 37, 40, 43, 44, 53] usually adopt the framework in Algorithm 1. Based on the techniques used, they can be classified into the following three categories: straightforward computation [16, 35], optimization methods [5, 13, 26, 53], and probabilistic graphical model methods [10, 11, 22, 23, 29–31, 37, 40, 43, 44]. Next these three categories will be discussed, respectively.

- **Straightforward Computation** Some baseline methods directly estimate v_i^* ($1 \leq i \leq n$) based on V, without modeling each worker or task. For decision-making and single-label tasks, majority voting (MV) regards the truth of each task as the answer given by most workers; while for numeric tasks, Mean and Median are two baseline methods that regard the mean and median of workers' answers as the truth for each task.
- **Optimization** The basic idea of optimization methods is to set a self-defined optimization function that captures the relations between workers' qualities and tasks' truth, and then derive an iterative method to compute these two sets of parameters collectively. The differences among existing works [5, 26, 27, 53] are that they model workers' qualities differently and apply different optimization functions to capture the relations between the two sets of parameters.
- (1) **Worker Probability** PM [5, 27] models each worker's quality as a single value, and the optimization function is defined as

$$\min_{\{q^w\},\{v_i^*\}} f(\{q^w\}, \{v_i^*\}) = \sum_{w \in \mathcal{W}} q^w \cdot \sum_{t_i \in \mathcal{T}^w} d(v_i^w, v_i^*),$$

    where $\{q^w\}$ represents the set of all workers' qualities and similarly $\{v_i^*\}$ represents the set of all truth. It models a worker w's quality as $q^w \geq 0$, and $d(v_i^w, v_i^*) \geq 0$ defines the distance between worker's answer v_i^w and the truth v_i^*: the more similar v_i^w is to v_i^*, the lower the value of $d(v_i^w, v_i^*)$ is. Intuitively, to minimize $f(\{q^w\}, \{v_i^*\})$, a worker w's high quality q^w corresponds to a low value in $d(v_i^*, v_i^w)$, i.e., worker w's answer should be close to the truth. By capturing the intuitions, similar to Algorithm 1, PM [5, 27] develops an iterative approach, and in each iteration, it adopts the two steps as illustrated in Sect. 3.2.2.
- (2) **Worker Probability and Confidence** Different from above, CATD [26] considers both worker probability and confidence in modeling a worker's quality. To be specific, each worker w's quality is scaled up to a coefficient of $\mathcal{X}^2_{(0.975, |\mathcal{T}^w|)}$, i.e., the more tasks w has answered, the higher worker w's quality is scaled to. It develops an objective function, with the intuitions that a worker w who gives answers close to the truth and answers a plenty of tasks should have a high quality q^w. Similarly it adopts an iterative approach and iterates the two steps until convergence.
- (3) **Diverse Skills** Minimax [53] leverages the idea of minimax entropy [55]. To be specific, it models the diverse skills of a worker w across different

tasks and focuses on single-label tasks (with ℓ choices). It assumes that for a task t_i, the answers given by w are generated by a probability distribution $\pi_{i,\cdot}^{w} = [\ \pi_{i,1}^{w}, \pi_{i,2}^{w}, \ldots, \pi_{i,\ell}^{w}\ ]$, where each $\pi_{i,j}^{w}$ is the probability that worker w answers task t_i with the j-th choice. Following this, an objective function is defined by considering two constraints for tasks and workers: for a task t_i, the number of answers collected for a choice equals the sum of corresponding generated probabilities; for a worker w, among all tasks answered by w, given the truth is the j-th choice, the number of answers collected for the k-th choice equals the sum of corresponding generated probabilities. Finally [53] devises an iterative approach to infer the two sets of parameters $\{v_i^*\}$ and $\{\pi^w\}$. There are also works [13, 31, 50] that consider the diverse qualities in a worker. To be specific, each worker is modeled as a quality vector of multiple domains, and each element in the quality vector captures the worker's quality for a specific domain.

- **Probabilistic Graphical Model (PGM)** A probabilistic graphical model [24] is a graph which expresses the conditional dependency structure (represented by edges) between random variables (represented by nodes). Figure 3.2 shows the general PGM adopted in existing works. Each node represents a variable. There are two plates, respectively, for workers and tasks, where each one represents the repeating variables. For example, the plate for workers represents $|\mathcal{W}|$ repeating variables, where each variable corresponds to a worker $w \in \mathcal{W}$. For the variables, α, β, and v_i^w are known (α and β are priors for q^w and v_i^*, which can be set based on the prior knowledge); q^w and v_i^* are latent or unknown variables, which are two desired variables to compute. The directed edges model the conditional dependence between a child node and its associated parent node(s) in the sense that the child node follows a probabilistic distribution conditioned on the values taken by the parent node(s). For example, three conditional distributions in Fig. 3.2 are $\Pr(q^w \mid \alpha)$, $\Pr(v_i^* \mid \beta)$, and $\Pr(v_i^w \mid q^w, v_i^*)$.

  Next we illustrate the details (optimization goal and the two steps) of each method using PGM. In general the methods differ in the used worker model. The methods can be classified into three categories: **worker probability** [11, 22, 29, 44], **confusion matrix** [10, 23, 37, 40], and **diverse skills** [13, 14, 31, 43]. For each category, we first introduce its basic method, e.g., ZC [11], and then summarize how other methods [22, 29, 44] extend the basic method ZC [11].

(1) Worker Probability: ZC [11] and Its Extensions [22, 29, 44]

ZC [11] adopts a PGM similar to Fig. 3.2, with the simplification that it does not consider the priors (i.e., α, β). Suppose all tasks are decision-making tasks ($v_i^* \in \{$T, F$\}$) and each worker's quality is modeled as worker probability $q^w \in [0, 1]$. Then we can derive

$$\Pr(v_i^w \mid q^w, v_i^*) = (q^w)^{\mathbb{1}\{v_i^w = v_i^*\}} \cdot (1 - q^w)^{\mathbb{1}\{v_i^w \neq v_i^*\}},$$

Fig. 3.2 The probabilistic
graphical model framework

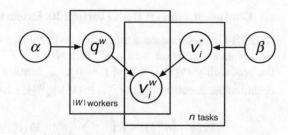

which means that the probability worker w correctly (incorrectly) answers a task is
$q^w (1 - q^w)$. For decision-making tasks, ZC [11] tries to maximize the probability
of the occurrence of workers' answers, called *likelihood*, i.e., $\max_{\{q^w\}} \Pr(V \mid \{q^w\})$,
which regards $\{v_i^*\}$ as latent variables:

$$\Pr(V \mid \{q^w\}) = \frac{1}{2} \cdot \prod_{i=1}^{n} \sum_{z \in \{T, F\}} \prod_{w \in \mathcal{W}^i} \Pr(v_i^w \mid q^w, v_i^* = z). \tag{3.1}$$

However, it is hard to optimize due to the non-convexity. Thus ZC [11] applies the
EM (expectation-maximization) framework [12] and iteratively updates q^w and v_i^*
to approximate its optimal value. Note ZC [11] develops a system to address entity
linking for online pages. This section focuses on leveraging the crowd's answers to
infer the truth (i.e., Section 4.3 in [11]), and we omit other parts (e.g., constraints on
its probabilistic model).

There are several extensions of ZC, e.g., GLAD [44], KOS [22], VI-BP [29], and
VI-MF [29], and they focus on different perspectives:

Task Model GLAD [44] extends ZC [11] in the task model. Rather than assuming
that each task is the same, it [44] models each task t_i's difficulty $d_i \in (0, +\infty)$ (the
higher, the easier). Then it models the worker's answer as $\Pr(v_i^w = v_i^* \mid d_i, q^w) =$
$1/(1 + e^{-d_i \cdot q^w})$ and integrates it into Eq. (3.1) to approximate the optimal value
using gradient descent [24] (an iterative method).

Optimization Function KOS [22], VI-BP [29], and VI-MF [29] extend ZC [11]
in the optimization goal. Recall that ZC tries to compute the optimal $\{q^w\}$ that
maximizes $\Pr(V \mid \{q^w\})$, which is the *point estimate*. Instead, [22, 29] leverage
the *Bayesian Estimators* to calculate the integral of all possible q^w, and the target is
to estimate the truth $v_i^* = \mathrm{argmax}_{z \in \{T, F\}} \Pr(v_i^* = z \mid V)$, where

$$\Pr(v_i^* = z \mid V) = \int_{\{q^w\}} \Pr(v_i^* = z, \{q^w\} \mid V) \, \mathrm{d}\{q^w\}. \tag{3.2}$$

It is hard to directly compute Eq. (3.2), and existing works [22, 29] seek for
variational inference (*VI*) techniques [42] to approximate the value: KOS [22] first
leverages *belief propagation* (one typical *VI* technique) to iteratively approximate
the value in Eq. (3.2), and then [29] proposes a more general model based on KOS,
called VI-BP. Moreover, it [29] also applies *mean field* (another *VI* technique) in
VI-MF to iteratively approach Eq. (3.2).

(2) Confusion Matrix: D&S [10] and Its Extensions [23, 37, 40]

D&S [10] focuses on single-label tasks (with fixed ℓ choices) and models each worker as a confusion matrix q^w with size $\ell \times \ell$. The worker w's answer follows the probability $\Pr(v_i^w \mid q^w, v_i^*) = q_{v_i^*, v_i^w}^w$. Similar to Eq. (3.1), D&S [10] tries to optimize the function $\mathrm{argmax}_{\{q^w\}} \Pr(V \mid \{q^w\})$, where

$$\Pr(V \mid \{q^w\}) = \prod_{i=1}^{n} \sum_{1 \le z \le \ell} \Pr(v_i^* = z) \cdot \prod_{w \in \mathcal{W}^i} q_{z, v_i^w}^w,$$

and it applies the EM framework [12] to devise two iterative steps.

The above method D&S [10], which models a worker as a confusion matrix, is also a widely used model. There are some extensions, e.g., LFC [37], LFC_N [37], BCC [23], and CBCC [40].

Priors LFC [37] extends D&S [10] to incorporate the priors into a worker's model, by assuming that the priors, denoted as $\alpha_{j,k}^w$ for $1 \le j, k \le \ell$, are known in advance, and it assumes that the worker's quality $q_{j,k}^w$ is generated following the Beta$(\alpha_{j,k}^w, \sum_{k=1}^{\ell} \alpha_{j,k}^w)$ distribution.

Task Type LFC_N [37] also handles numeric tasks. Different from decision-making and single-choice tasks, it assumes that worker w's answer follows $v_i^w \sim \mathcal{N}(v_i^*, \sigma_w^2)$, where σ_w is the variance, and a small σ_w implies that v_i^w is close to the truth v_i^*.

Optimization Function BCC [23] has a different optimization goal compared with D&S [10], and it aims at maximizing the posterior joint probability. For example, in Fig. 3.2, it optimizes the posterior joint probability of all unknown variables, i.e.,

$$\prod_{i=1}^{n} \Pr(v_i^* \mid \beta) \prod_{w \in \mathcal{W}} \Pr(q^w \mid \alpha) \prod_{i=1}^{n} \prod_{w \in \mathcal{W}^i} \Pr(v_i^w \mid q^w, v_i^*).$$

To optimize the above formula, the technique of Gibbs sampling [24] is used to iteratively infer the two sets of parameters $\{q^w\}, \{v_i^*\}$ until convergence, where q^w is modeled as a confusion matrix. Then CBCC [40] extends BCC [23] to support community. The basic idea is that each worker belongs to one community, where each community has a representative confusion matrix, and workers in the same community share very similar confusion matrices.

(3) Diverse Skills: Multi [43] and Others [13, 31, 50]

Recently, there are some works (e.g., [13, 31, 43, 50]) that model a worker's diverse skills. Basically, they model a worker w's quality q^w as a vector of size K, which captures a worker's diverse skills over K latent topics. For example, [31] combines the process of topic modeling (i.e., TwitterLDA [46]) and truth inference together, and [50] leverages entity linking and knowledge bases to exploit a worker's diverse skills.

This section summarizes the similarities (Sect. 3.2.2) and differences (Sects. 3.2.3.1 3.2.3.2 and 3.2.3.3) of existing works. There is a recent paper [51] that evaluates the performance of different existing truth inference solutions in practice. The authors have also made all the codes and datasets public. Interested readers can refer to [51] for more detailed analysis.

3.2.4 Extensions of Truth Inference

Based on the above discussions on truth inference, there are some works that leverage the truth inference techniques in quality control.

- **Worker Elimination** Worker elimination, which eliminates low-quality workers or spammers based on worker models, is a very common strategy to improve quality [20, 32, 36]. There are different ways to detect low-quality workers or spammers.

  A simple way is to use **qualification test** [1] or **hidden test** [3] as introduced above. For example, before a worker answers real tasks, a qualification test that consists of golden tasks should be accomplished by the worker. Based on the worker's answers on the golden tasks, the worker's quality can be computed. Then, the workers with low qualities (e.g., with worker probability <0.6) will be blocked to answer real tasks.

  Other than the worker probability model, there are some spammer detection methods based on more complex worker models. For example, Raykar et al. [36] and Ipeirotis et al. [20] study how to detect spammers if each worker is modeled as a confusion matrix (an $\ell \times \ell$ matrix). The basic idea is that for each worker, they compute a score based on the worker's confusion matrix, and the score represents how indicative the worker's answer is on a task's true answer. The higher the worker's score is, the more indicative her/his answer is on a task's true answer, and the more reliable s/he is. Then, they block the workers whose scores are below a predefined threshold.

  In addition to detecting spammers based on worker models, Marcus et al. [32] propose a detection method that is based on inconsistent answers given by different workers. They first compute the deviation of a worker's answer from the majority of other workers' answers. If the worker's answer deviates a lot from the majority of other workers' answers, the worker will be blocked. Yuan [45] et al. studied how to detect sybil workers that intentionally give wrong errors by simulating multiple workers and submitting the answers to dominate the majority of the answers.

- **Incremental Truth Inference** The above solutions to truth inference are iterative methods, while in many cases workers' answers come at a high velocity, and an incremental method is needed, which instantly updates previously stored parameters when a new answer arrives. The basic idea of [15, 50] is that upon receiving a worker's new answer, it only chooses a small subset of related parameters to update, e.g., the truth of answered task, and the models of workers who have answered the task before. In practice, the incremental approach can be integrated into the iterative approach in a delayed manner (in order to improve the efficiency), i.e., the iterative approach is run when receiving every k (e.g., $k = 100$) answers; and among the k answers, the parameters are updated using the incremental approach.

3.3   Task Assignment

To formulate the problem of crowdsourced task assignment, existing works focused on two scenarios: (1) task assignment setting, i.e., when a worker comes, which subset of tasks should be assigned to the worker, and (2) worker selection setting, i.e., given a task, which subset of workers should be selected to answer the task. We next analyze these two settings in Sects 3.3.1 and 3.3.2, respectively.

3.3.1   Task Assignment Setting

In the task assignment setting, when a worker comes, existing works [7, 13, 30, 52] address an *online task assignment problem*, which is defined as follows:

Definition 3.9 (Online Task Assignment Problem) Given a pool of n tasks, when a worker comes to answer tasks, which set of the k tasks should be batched in a Human Intelligence Task (HIT) and assigned to the coming worker?

Figure 3.3 gives an example of the problem. Suppose we have $n = 4$ tasks, and each HIT contains $k = 2$ tasks. When a worker comes to answer tasks, we are interested in which two-task combination should be batched in a HIT and assigned to the coming worker.

Fig. 3.3 Online task assignment problem

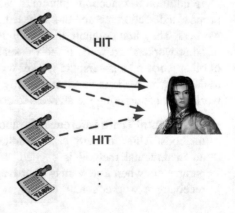

In AMT, this issue is usually addressed in an *offline* manner: the tasks assigned to all HITs are decided before they are shown to the workers. As pointed out in [7, 30], the main drawback of this approach is that the difficulty level of a task is not considered: for an "easy" task, its final result can be determined even if the current number of answers received from workers is limited, whereas a more difficult or controversial task may require more answers from more workers. Notice that a requester may only have a limited amount of budget to pay the workers. It is thus important to select the tasks that are batched in a HIT, in order to obtain the best answers under the limited budget. Moreover, the problem is inherently complex, since finding the best solution for the task assignment problem can be extremely expensive. Given a pool of n tasks, there are $\binom{n}{k}$ sets of candidate tasks for a HIT. Fast assignment response to the worker is required as the worker may feel bored if waiting for a long time.

The problem is complex since a simple enumeration would consider all $\binom{n}{k}$ combinations. Consider $n = 100$ and $k = 5$; then simple enumeration would consider all $100M$ possible combinations of assignments, which is computationally expensive. Moreover, there requires efficient (online) task assignment, since the problem requires fast response to worker's request. A reasonable assignment time is that the assignment process can be finished within $\mathcal{O}(n)$ [25], i.e., linear to the number of tasks.

To address the problem, existing works [7, 30] often focus on assigning the most suitable tasks to a worker. To define the notion "suitable," existing works [7, 13, 17, 18, 30, 38, 50] consider the following three factors:

Answer Uncertainty, which analyzes the uncertainty of answers based on the collected answers from multiple workers.

Worker Quality, which considers the coming worker's quality and assigns tasks to the worker with suitable expertise.

Objectives of Requesters, which assigns tasks to a worker by considering the specified evaluation metric.

We have summarized the three factors that different methods considered in Table 3.5. Next, we discuss these three factors in Sects. 3.3.1.1, 3.3.1.2, and 3.3.1.3, respectively.

3.3.1.1 Answer Uncertainty

For answer uncertainty, given the collected answers, exiting works [7, 30, 52] select a task whose answers are the most *uncertain* or *inconsistent*. Suppose we have four tasks t_1, t_2, t_3, and t_4, and each task asks workers to answer "*Yes*" or "*No*," and each task receives three answers. Let (n_{Yes}, n_{No}) denote the number of Yes's and No's collected, and t_1, t_2, t_3, and t_4 receive $(3, 0)$, $(2, 1)$, $(1, 2)$, and $(0, 3)$ as the answers, respectively. Then when a worker comes, it should be more reasonable to assign t_2 and t_3 to the coming worker, since it is more beneficial to assign the tasks with inconsistent answers.

Table 3.5 Comparisons of task assignment methods (task selection)

Method	Answer uncertainty	Worker quality	Objectives of requesters
OTA [18]	Restriction on #collected answers	Worker probability	Maximize total utility
CDAS [30]	Heuristics	Worker probability	Early termination
iCrowd [13]	Heuristics	Diverse domains	Maximize overall Worker quality
AskIt! [7]	Expected change of entropy	No	No
QASCA [52]	Heuristics	Confusion matrix	Maximize specified quality
DOCS [50]	Expected change of entropy	Diverse domains	Maximize the benefit
CrowdPOI [19]	Expected change of accuracy	Worker probability	Maximize the benefit
Opt-KG [28]	Heuristics	No	Threshold on entropy

By leveraging this idea, existing works propose different ways to define the answer uncertainty. For example, the entropy of the collected answers of a take can be defined as the answer uncertainty of the task. Then the tasks with the highest entropy will be assigned to a coming worker [52]. For example, given c choices for a task, and the probability distribution of each choice inferred from the task's answers is denoted as $\mathbf{p} = (p_1, p_2, \ldots, p_c)$, then the entropy for the task is computed as $H(\mathbf{p}) = -\sum_{i=1}^{c} p_i \cdot \log p_i$. Consider that a decision-making task receives one "*Yes*" and two "*No*" as answers; then a naive way to derive the distribution is to calculate the proportion of answers received for each choice, i.e., $(1/3, 2/3)$, and the entropy is 0.637.

Besides entropy, the expected improvement of entropy [7, 50] is used in task assignment, where entropy improvement is $H(\mathbf{p}^{t,w}) - H(\mathbf{p})$, in which $H(\mathbf{p}^{t,w})$ is the entropy computed by assuming that the task t is answered by working w and $H(\mathbf{p})$ is the entropy that t is not answered by w. Then given a coming worker, the method assigns the task with the highest improvement in entropy. Since it is hard to compute $H(\mathbf{p}^{t,w})$ if the task t is not answered by worker w, existing works [7, 50] estimate the expected change of entropy. The intuition is that consider two decision-making tasks: the first receives 1 *Yes* and 2 *No*, while the second receives 10 *Yes* and 20 *No* as answers. If one only considers entropy, then the entropy of the two tasks are the same. However, since the first task receives much less answers compared to the second one, the benefit of assigning the task (measured by the improvement of entropy) should be much higher than that of the second one. Similar ideas were used in [19], which leverages the expected improvement of accuracy to estimate the answer uncertainty.

Also, there are some works [18] that restricts the number of assignments for each task. The basic idea is that for hard tasks, they should be assigned to more workers, while for easy tasks, they should be assigned to less workers. Some other works [13, 28, 30, 52] proposed heuristics to define the answer uncertainty.

3.3.1.2 Worker Quality

Another factor is to consider the worker's expertise, and assign tasks to the worker with suitable expertise. Many existing approaches [18, 19, 30, 52] model the quality of each worker as either worker probability or a confusion matrix. Some recent works [13, 50] model a worker's diverse skills among different domains, and when a worker comes to request tasks, the method selects tasks with the matching domains to the worker. For example, if the coming worker has high quality in domain *sports* and there are three candidate tasks t_1, t_2, and t_3, which are related to *politics*, *entertainment*, and *sports*, respectively, then it will be more beneficial to assign task t_3 to the coming worker, since the task has matching expertise (w.r.t. domain *sports*) with the worker.

Note that the assignment requires worker's quality to be known in advance. There are two cases to consider: (1) if the worker has answered tasks before, then truth inference methods (discussed in Sect. 3.2) can help to compute each worker's quality (as shown in the truth inference components in Fig. 3.1); (2) if the worker is a new worker, then the worker's quality can be initialized as the average quality of all past workers [50].

3.3.1.3 Objectives of Requesters

There are some recent works [13, 18, 28, 52] that consider the specific objectives of requesters. For example, Zheng et al. [52] find that different crowdsourcing applications may have different ways to define quality. In their approach, the requester will first specify a quality metric (e.g., accuracy, F-score) that aims to be optimized over the data. To meet the requirement, the assignment algorithm will decide which k tasks should be assigned according to the specified metric. For each combination of k tasks, it computes how much the quality will be improved if the tasks are assigned to a coming worker and selects the combination that can lead to the maximum improvement in quality.

To be specific, [52] models all questions using a distribution matrix Q^c (with size $n \times \ell$), where each row $Q_i^c = [\ Q_{i,1}^c, Q_{i,2}^c, \ldots, Q_{i,\ell}^c\ ]$ represents the probability of the i-th task's true answers (computed based on the current collected answers). Then when a worker comes, based on the worker's quality (can be learned from the worker's past answering performance), for each k-task combination, a new distribution matrix Q^X (with size $n \times \ell$) is computed, where each row $Q_i^X = [\ Q_{i,1}^X, Q_{i,2}^X, \ldots, Q_{i,\ell}^X\ ]$ represents the probability of the i-th task's true answers if the coming worker answers the selected k tasks. To capture the goodness of a given distribution matrix, [52] introduces a function $f(\cdot)$ which takes the distribution matrix as an input, and outputs a score indicating the quality of the matrix. The function $f(\cdot)$ takes the requester-specified evaluation metric into consideration and [52] develops two functions, respectively, for accuracy and F-score. Then the task assignment problem is formalized to select the optimal k-task combination (denoted as X^*), such that the quality of improvement w.r.t. the specified evaluation metric can be maximized, i.e., $X^* = \arg\max_X \{f(Q^X) - f(Q^c)\}$. Intuitively, [52] considers worker's quality in the assignment and examines how the quality (defined on specified evaluation metric) of all tasks may change for each k-task combination and then selects the optimal combination that can maximize the improvement in quality.

Note that there are also other works [13, 18, 28] that consider different objectives. For example, [28] sets a threshold on the entropy (e.g., 0.6), and it requires that in the final state, each task should have the constraint that each task's entropy (computed based on the collected answers) should be ≥ 0.6. Similarly, [13] sets a threshold on the worker quality (e.g., 2.0), and it requires that in the final state, each task should have the overall aggregated worker quality ≥ 2.0. Ho and Vaughan [18] aims to maximize the total utility, and after an answer is given by some workers, the requester will receive some utility related to worker quality. The objective is to assign tasks that maximize the total utility. Liu et al. [30] applies an early termination strategy, and the basic idea is to terminate assigning very confident tasks to workers. Also, there are other works [19, 50] that proposes heuristics to maximize the benefit of assignment.

3.3.2 Worker Selection Setting

In the worker selection setting, given a task and a set of workers (with known qualities), intuitively the workers with high qualities (or having matching skills [47, 48] to the task) should be selected. In addition to these factors, *worker cost* is another key factor for the worker selection [9, 49], which is the monetary cost that each worker requires to answer a task. The cost can be indicated by the worker or learned from the worker's profiles [9] (e.g., registration date, academic degree).

Considering the given worker budget, Cao et al. [9] propose the *jury selection problem* (JSP):

Definition 3.10 (Jury Selection Problem) Given a task, a set of workers (with known qualities and costs), and an overall budget, how are we going to select a subset of workers in order to maximize the task's quality without exceeding the overall budget?

Figure 3.4 shows a decision-making task, to be answered by some of the seven workers labeled from A to G where each worker is associated with a *quality* and a *cost*. The *quality* ranges from 0 to 1, indicating the probability that the worker correctly answers a question. This probability can be estimated by using her background information (see Sect. 3.2). The *cost* is the amount of monetary reward the worker can get upon finishing a task. In this example, A has a quality of 0.77 and a cost of 9 units. For a jury (i.e., a set of selected workers), the *jury cost* is defined as the sum of workers' costs in the jury and the *jury quality* (or JQ) is defined as the probability that the result returned by aggregating the jury answers is correct.

Fig. 3.4 Jury selection example

Table 3.6 Comparisons of task assignment methods (worker selection)

Method	Jury quality computation	Jury selection problem
Cao et al. [9]	Majority voting	NP-hard
	Exact solution with $\mathcal{O}(|S| \cdot \log |S|)$ time	A polynomial heuristic algorithm
Zheng et al. [49]	Bayesian voting	NP-hard
	Approximate solution with $\mathcal{O}(|S|^3)$ time	Simulated annealing heuristic

Given a budget of B units, a feasible jury is a jury whose *jury cost* does not exceed B. For example, if $B = \$20$, then $\{B, E, F\}$ is a feasible jury, since its *jury cost*, i.e., $\$5 + \$5 + \$2 = \$12$, is not larger than $\$20$.

To solve JSP, a naive solution is to compute the JQ for every feasible jury, and return the one with the highest JQ. In [9], the authors study how to compute JQ for a jury where the jury's returned result is decided by *majority voting* (MV; see Sect. 3.2). In the following, we consider each worker's answer as a "vote" for either "Yes" or "No." Let us consider $\{B, E, F\}$ again, the probability that these workers give a correct result according to MV is $0.7 \cdot 0.6 \cdot 0.6 + 0.7 \cdot 0.6 \cdot (1 - 0.6) + 0.7 \cdot (1 - 0.6) \cdot 0.6 + (1 - 0.7) \cdot 0.6 \cdot 0.6 = 69.6\%$. Moreover, since $\{A, C, G\}$ yields the highest JQ among all the feasible juries, then it is considered as the optimal result by [9]. Moreover, [9] proposes an algorithm with time complexity of $\mathcal{O}(|S| \cdot \log|S|)$, where S is the set of given workers. Finally, [9] proves that solving JSP is NP-hard, and proposes a polynomial heuristic algorithm to decide the jury that should be selected.

Recently, Zheng et al. [49] prove that Bayesian Voting is the optimal strategy under the definition of JQ. That is, given any fixed S, the JQ of S w.r.t. the Bayesian Voting strategy is not lower than the JQ of S w.r.t. any other strategy. So given a set of workers, its collective quality (or JQ) w.r.t. Bayesian Voting strategy is the highest among all voting strategies. Zheng et al. [49] further proves that the computation of JQ w.r.t. the Bayesian Voting strategy is NP-hard. To reduce the computational complexity, they propose an $\mathcal{O}(|S|^3)$ approximation algorithm, within 1% error bound. Also, [49] proves that solving the optimal jury selection problem is NP-hard and leverages the simulated annealing heuristic to decide the jury that should be selected.

Table 3.6 summarizes the characteristics of the two approaches.

3.4 Summary of Quality Control

This chapter reviewed the techniques of quality control in crowdsourcing and has especially focused on two important problems: (1) Truth inference, which collects workers' answers and infers the truth of each task by leveraging all collected answers from workers, and (2) task assignment, which assigns tasks to appropriate workers.

In truth inference (Sect. 3.2), the problem was formally defined (Sect. 3.2.1), and then the similarities and differences in existing works were summarized. Specifically, a unified solution framework adopted in existing works was introduced (Sect. 3.2.2), and different factors considered in various works were discussed (Sect. 3.2.3). Finally, several extensions of truth inference were discussed (Sect. 3.2.4).

In task assignment (Sect. 3.3), we focused on two settings adopted in existing works: task assignment setting and worker selection setting. In task assignment setting (Sect. 3.3.1), this chapter summarized different factors (e.g., answer uncertainty, worker quality, requester requirement) that have been considered in existing works; in worker selection setting (Sect. 3.3.2), the chapter discussed the jury selection problem and various solutions of this problem.

References

1. Amazon mechanical turk. https://www.mturk.com/
2. Chi-squared distribution. https://en.wikipedia.org/wiki/Chi-squared_distribution
3. Crowdflower. http://www.crowdflower.com
4. External hit. http://docs.aws.amazon.com/AWSMechTurk/latest/AWSMturkAPI/Welcome.html
5. Aydin, B.I., Yilmaz, Y.S., Li, Y., Li, Q., Gao, J., Demirbas, M.: Crowdsourcing for multiple-choice question answering. In: AAAI, pp. 2946–2953 (2014)
6. Blei, D.M., Ng, A.Y., Jordan, M.I.: Latent dirichlet allocation. JMLR 3(Jan), 993–1022 (2003)
7. Boim, R., Greenshpan, O., Milo, T., Novgorodov, S., Polyzotis, N., Tan, W.C.: Asking the right questions in crowd data sourcing. In: ICDE, pp. 1261–1264 (2012)
8. Callison-Burch, C.: Fast, cheap, and creative: evaluating translation quality using amazon's mechanical turk. In: EMNLP, pp. 286–295 (2009)
9. Cao, C.C., She, J., Tong, Y., Chen, L.: Whom to ask? jury selection for decision making tasks on micro-blog services. PVLDB 5(11), 1495–1506 (2012)
10. Dawid, A.P., Skene, A.M.: Maximum likelihood estimation of observer error-rates using the em algorithm. Applied statistics pp. 20–28 (1979)
11. Demartini, G., Difallah, D.E., Cudré-Mauroux, P.: Zencrowd: leveraging probabilistic reasoning and crowdsourcing techniques for large-scale entity linking. In: WWW, pp. 469–478 (2012)
12. Dempster, A.P., Laird, N.M., Rubin, D.B.: Maximum likelihood from incomplete data via the em algorithm. J.R.Statist.Soc.B 30(1), 1–38 (1977)
13. Fan, J., Li, G., Ooi, B.C., Tan, K., Feng, J.: icrowd: An adaptive crowdsourcing framework. In: SIGMOD, pp. 1015–1030 (2015)
14. Fang, Y., Sun, H., Li, G., Zhang, R., Huai, J.: Effective result inference for context-sensitive tasks in crowdsourcing. In: DASFAA, pp. 33–48 (2016)

15. Feng, J., Li, G., Wang, H., Feng, J.: Incremental quality inference in crowdsourcing. In: DASFAA, pp. 453–467 (2014)
16. Franklin, M.J., Kossmann, D., Kraska, T., Ramesh, S., Xin, R.: Crowddb: answering queries with crowdsourcing. In: SIGMOD, pp. 61–72 (2011)
17. Ho, C.J., Jabbari, S., Vaughan, J.W.: Adaptive task assignment for crowdsourced classification. In: ICML, pp. 534–542 (2013)
18. Ho, C.J., Vaughan, J.W.: Online task assignment in crowdsourcing markets. In: AAAI (2012)
19. Hu, H., Zheng, Y., Bao, Z., Li, G., Feng, J.: Crowdsourced poi labelling: Location-aware result inference and task assignment. In: ICDE, pp. 61–72 (2016)
20. Ipeirotis, P., Provost, F., Wang, J.: Quality management on amazon mechanical turk. In: SIGKDD Workshop, pp. 64–67 (2010)
21. Joglekar, M., Garcia-Molina, H., Parameswaran, A.G.: Evaluating the crowd with confidence. In: SIGKDD, pp. 686–694 (2013)
22. Karger, D.R., Oh, S., Shah, D.: Iterative learning for reliable crowdsourcing systems. In: NIPS, pp. 1953–1961 (2011)
23. Kim, H.C., Ghahramani, Z.: Bayesian classifier combination. In: AISTATS, pp. 619–627 (2012)
24. Koller, D., Friedman, N.: Probabilistic Graphical Models - Principles and Techniques. MIT Press (2009)
25. Li, G., Zheng, Y., Fan, J., Wang, J., Cheng, R.: Crowdsourced data management: Overview and challenges. In: SIGMOD, pp. 1711–1716 (2017)
26. Li, Q., Li, Y., Gao, J., Su, L., Zhao, B., Demirbas, M., Fan, W., Han, J.: A confidence-aware approach for truth discovery on long-tail data. PVLDB 8(4), 425–436 (2014)
27. Li, Q., Li, Y., Gao, J., Zhao, B., Fan, W., Han, J.: Resolving conflicts in heterogeneous data by truth discovery and source reliability estimation. In: SIGMOD, pp. 1187–1198 (2014)
28. Li, Q., Ma, F., Gao, J., Su, L., Quinn, C.J.: Crowdsourcing high quality labels with a tight budget. In: WSDM, pp. 237–246 (2016)
29. Liu, Q., Peng, J., Ihler, A.T.: Variational inference for crowdsourcing. In: NIPS, pp. 701–709 (2012)
30. Liu, X., Lu, M., Ooi, B.C., Shen, Y., Wu, S., Zhang, M.: CDAS: A crowdsourcing data analytics system. PVLDB 5(10), 1040–1051 (2012)
31. Ma, F., Li, Y., Li, Q., Qiu, M., Gao, J., Zhi, S., Su, L., Zhao, B., Ji, H., Han, J.: Faitcrowd: Fine grained truth discovery for crowdsourced data aggregation. In: KDD, pp. 745–754 (2015)
32. Marcus, A., Karger, D.R., Madden, S., Miller, R., Oh, S.: Counting with the crowd. PVLDB 6(2), 109–120 (2012)
33. Marcus, A., Wu, E., Madden, S., Miller, R.C.: Crowdsourced databases: Query processing with people. In: CIDR, pp. 211–214 (2011)
34. Parameswaran, A.G., Garcia-Molina, H., Park, H., Polyzotis, N., Ramesh, A., Widom, J.: Crowdscreen: algorithms for filtering data with humans. In: SIGMOD, pp. 361–372 (2012)
35. Parameswaran, A.G., Park, H., Garcia-Molina, H., Polyzotis, N., Widom, J.: Deco: declarative crowdsourcing. In: CIKM, pp. 1203–1212. ACM (2012)
36. Raykar, V.C., Yu, S.: Eliminating spammers and ranking annotators for crowdsourced labeling tasks. Journal of Machine Learning Research 13, 491–518 (2012)
37. Raykar, V.C., Yu, S., Zhao, L.H., Valadez, G.H., Florin, C., Bogoni, L., Moy, L.: Learning from crowds. JMLR 11(Apr), 1297–1322 (2010)
38. Roy, S.B., Lykourentzou, I., Thirumuruganathan, S., Amer-Yahia, S., Das, G.: Task assignment optimization in knowledge-intensive crowdsourcing. VLDBJ 24(4), 467–491 (2015)
39. Shen, W., Wang, J., Han, J.: Entity linking with a knowledge base: Issues, techniques, and solutions. TKDE 27(2), 443–460 (2015)
40. Venanzi, M., Guiver, J., Kazai, G., Kohli, P., Shokouhi, M.: Community-based bayesian aggregation models for crowdsourcing. In: WWW, pp. 155–164 (2014)
41. Von Ahn, L., Maurer, B., McMillen, C., Abraham, D., Blum, M.: recaptcha: Human-based character recognition via web security measures. Science 321(5895), 1465–1468 (2008)

42. Wainwright, M.J., Jordan, M.I.: Graphical models, exponential families, and variational inference. Foundations and Trends in Machine Learning **1**(1–2), 1–305 (2008)
43. Welinder, P., Branson, S., Perona, P., Belongie, S.J.: The multidimensional wisdom of crowds. In: NIPS, pp. 2424–2432 (2010)
44. Whitehill, J., Ruvolo, P., Wu, T., Bergsma, J., Movellan, J.R.: Whose vote should count more: Optimal integration of labels from labelers of unknown expertise. In: NIPS, pp. 2035–2043 (2009)
45. Yuan, D., Li, G., Li, Q., Zheng, Y.: Sybil defense in crowdsourcing platforms. In: CIKM, pp. 1529–1538 (2017)
46. Zhao, W.X., Jiang, J., Weng, J., He, J., Lim, E.P., Yan, H., Li, X.: Comparing twitter and traditional media using topic models. In: ECIR, pp. 338–349 (2011)
47. Zhao, Z., Wei, F., Zhou, M., Chen, W., Ng, W.: Crowd-selection query processing in crowdsourcing databases: A task-driven approach. In: EDBT, pp. 397–408 (2015)
48. Zhao, Z., Yan, D., Ng, W., Gao, S.: A transfer learning based framework of crowd-selection on twitter. In: SIGKDD, pp. 1514–1517 (2013)
49. Zheng, Y., Cheng, R., Maniu, S., Mo, L.: On optimality of jury selection in crowdsourcing. In: EDBT, pp. 193–204 (2015)
50. Zheng, Y., Li, G., Cheng, R.: DOCS: domain-aware crowdsourcing system. PVLDB **10**(4), 361–372 (2016)
51. Zheng, Y., Li, G., Li, Y., Shan, C., Cheng, R.: Truth inference in crowdsourcing: Is the problem solved? PVLDB **10**(5), 541–552 (2017)
52. Zheng, Y., Wang, J., Li, G., Cheng, R., Feng, J.: QASCA: A quality-aware task assignment system for crowdsourcing applications. In: SIGMOD, pp. 1031–1046 (2015)
53. Zhou, D., Basu, S., Mao, Y., Platt, J.C.: Learning from the wisdom of crowds by minimax entropy. In: NIPS, pp. 2195–2203 (2012)
54. Zhou, D., Liu, Q., Platt, J., Meek, C.: Aggregating ordinal labels from crowds by minimax conditional entropy. In: ICML, pp. 262–270 (2014)
55. Zhu, S., Wu, Y., Mumford, D.: Minimax entropy principle and its application to texture modeling. Neural computation **9**(8), 1627–1660 (1997)

Chapter 4
Cost Control

Despite the availability of crowdsourcing platforms, which provide a much cheaper way to ask humans to do some work, it is still quite expensive when there is a lot of work to do. Therefore, a big challenge in crowdsourced data management is cost control, i.e., how to reduce human cost while still keeping good result quality.

This chapter first presents the basic ideas of cost control in Sect. 4.1. After that, it discusses five classes of techniques to achieve this goal. Task pruning studies how to prune unnecessary tasks (Sect. 4.2). Answer deduction studies how to deduce crowd answers without asking the crowd to actually do the tasks (Sect. 4.3). Task selection studies how to select the most beneficial tasks (Sect. 4.4). Sampling studies how to ask the crowd to work on a sample of tasks (Sect. 4.5). Task design studies how to optimize task interface (Sect. 4.6).

4.1 Overview of Cost Control

The cost of completing a crowdsourcing job can be roughly written in the following form:

$$\text{cost} = n \cdot c,$$

where n denotes the total number of tasks and c denotes the monetary cost to complete each task. For example, suppose a requester wants to label 100 images. She can create 100 tasks where each task is to pay a crowd worker \$0.1 to label one image. Thus, the total cost is

$$\text{cost} = 100 * \$0.1 = \$10.$$

© Springer Nature Singapore Pte Ltd. 2018
G. Li et al., *Crowdsourced Data Management*,
https://doi.org/10.1007/978-981-10-7847-7_4

This formula suggests that cost control mainly has two basic ideas: (1) reducing the number of tasks and (2) reducing the cost of each task. The database community is mainly focused on the use of the first idea for cost control and studies how to reduce the number of tasks without any or only a little loss in answer quality. The second idea is the main focus of the HCI community, which aims to design a more user-friendly interface such that a crowd worker can finish a task with less effort. Next, this chapter will discuss five classes of techniques for reducing the number of tasks as well as their pros and cons.

4.2 Task Pruning

The first class of techniques is to use computer algorithms to preprocess all the tasks and then prune the tasks that do not need to be checked by humans. The underlying idea of this technique is that in many situations, there are a lot of tasks that can be easily finished by computers; thus humans only need to do the most challenging ones. In the following, we will take *crowdsourced join* as an example to illustrate this technique.

Definition 4.1 (Crowdsourced Join) Crowdsourced join (a.k.a. crowdsourced entity resolution) aims to find all pairs of records in a table (or between two tables) that refer to the same real-world entity.

For example, consider the table of product data shown in Table 4.1. Records r_1 and r_2 in the table have different text in the Product Name field but refer to the same product. The goal of crowdsourced join is to find all such matching records, i.e., (r_1, r_2), (r_1, r_7), (r_3, r_4), and (r_2, r_7).

For a table with n records, a naive way to solve the problem is to create $\frac{n(n-1)}{2}$ tasks, where each task is to ask the crowd to determine whether a pair of records refer to the same entity or not. This human-only approach is prohibitively expensive. Even with a modest table size of $n = 10{,}000$ records, this approach would require 49,995,000 tasks. At even \$0.01 per task, it would cost nearly half a million dollars to complete all the tasks.

Table 4.1 A table of products

ID	Product name	Price
r_1	iPad Two 16GB WiFi White	\$490
r_2	iPad 2nd generation 16GB WiFi White	\$469
r_3	iPhone 4th generation White 16GB	\$545
r_4	Apple iPhone 4 16GB White	\$520
r_5	Apple iPhone 3rd generation Black 16GB	\$375
r_6	iPhone 4 32GB White	\$599
r_7	Apple iPad2 16GB WiFi White	\$499
r_8	Apple iPod shuffle 2GB Blue	\$49
r_9	Apple iPod shuffle USB Cable	\$19

To reduce the cost, a key observation is that among $\frac{n(n-1)}{2}$ record pairs, most of them are quite easy for machines to do well. For example, consider $r_6 =$ "iPhone 4 32GB White" and $r_9 =$ "Apple iPod shuffle USB Cable." Since they do not share any word, a machine-based technique can easily identify the pair as a nonmatching pair with high confidence. Based on this idea, task pruning first measures the difficulty of each task and then prunes the tasks whose difficulties are smaller than a user-specified threshold. For the pruned tasks, trust machines' answers; for the remaining ones, utilize the crowd to get answers. To apply this technique, we need to address two challenges: (1) how to quantify the *difficulty* of each task and (2) how to specify a threshold to distinguish between easy tasks and difficult tasks.

4.2.1 Difficulty Measurement

Intuitively, if machines are not certain about a task's answer, it means that the task is difficult for machines to do correctly. Thus, answer uncertainty is a good measurement of task difficulty. Different machine-based entity resolution (ER) techniques have different ways to quantify answer uncertainty. We next introduce two categories of ER techniques: similarity-based and learning-based [5, 9].

Similarity-based Similarity-based techniques use a similarity function and a threshold to determine whether a pair of records is matching or not. Given a pair of records, they first compute the similarity of the two records. If the similarity value is larger than the threshold, it will be considered as a matching pair and otherwise a nonmatching pair. For example, in Table 4.1, suppose that the similarity of two records is specified as Jaccard similarity between their Product Names. Jaccard similarity over two sets is defined as the size of the set intersection divided by the size of the set union. For example, the Jaccard similarity between the Product Names of r_1 and r_2 is

$$\text{JAC}(r_1, r_2) = \frac{|\{\texttt{iPad, 16GB, WiFi, White}\}|}{|\{\texttt{iPad, 16GB, WiFi, White, Two, 2nd, generation}\}|} = 0.57.$$

If the threshold is set to 0.5, then r_1 and r_2 will be identified as matching since $\text{JAC}(r_1, r_2) = 0.57 \geq 0.5$. It is easy to see that the more close the similarity value is to 0 or 1, the more certain a similarity-based approach is to correctly identify its answer. Therefore, we derive the following equation to quantify task difficulty:

$$\text{DIFFICULTY}(r_1, r_2) = \min\{\text{JAC}(r_1, r_2), 1 - \text{JAC}(r_1, r_2)\},$$

which is the minimum of the differences of a similarity value from 0 and 1. For example, suppose $\text{JAC}(r_1, r_2) = 0.57$. Then, the difficulty of determining whether r_1 and r_2 are matching or not is $\text{DIFFICULTY}(r_1, r_2) = \min\{0.57, 1 - 0.57\} =$

0.43. Consider another record pair, r_6 and r_9. Since $\text{JAC}(r_6, r_9) = 0$, then we have $\text{DIFFICULTY}(r_6, r_9) = 0$; thus this pair is easier for machines to do correctly.

Learning-based Learning-based techniques model entity resolution as a classification problem. They represent a pair of records as a feature vector in which each dimension is a similarity value of the records on some attribute. If we choose n similarity functions on m attributes, then the feature vector will be a nm-dimensional feature vector. For example, for the records in Table 4.1, suppose we only choose Jaccard similarity on Product Name. Then each pair of records will be represented as a feature vector that contains only a single dimension. Learning-based techniques require a training set to train the classifier. The training set consists of positive feature vectors and negative feature vectors indicating matching pairs and nonmatching pairs, respectively. The trained classifier can then be applied to label new record pairs as matching or nonmatching. In addition, the classifier often outputs a confidence value for its prediction. A confidence value is in the range of $[0.5, 1]$. The closer the value is to 0.5, the more uncertain the classifier is to correctly identify its answer. Thus, we derive the following equation to quantify task difficulty:

$$\text{DIFFICULTY}(r_1, r_2) = 1 - \text{CONFIDENCE}\big(Label(r_1, r_2)\big),$$

where $Label(r_1, r_2)$ denotes the predicted label (matching or nonmatching) of a record pair and $\text{CONFIDENCE}(\cdot)$ represents the confidence for the predicted label.

4.2.2   Threshold Selection

A threshold is a parameter used to balance the trade-off between cost and quality. Having a larger threshold prunes more tasks, but on the other hand, since more difficult tasks are assigned to machines, it will potentially decrease overall quality.

Given a cost budget, we can infer how many tasks can be crowdsourced within the budget and then calculate the corresponding threshold. For example, suppose a user has a budget of \$50 and is willing to pay \$0.1 for each task. Then, we can infer that $\frac{\$50}{\$0.1} = 500$ tasks can be crowdsourced in total. We will set the threshold to a value such that the 500 most difficult tasks will be left after pruning.

Another idea is to plot the difficulty distribution of all tasks. An interesting observation is that this distribution often follows a power-law distribution. That is, the majority of tasks have very low difficulties and thus can be easily done by machines. Thus, we can select a relatively small threshold (e.g., 0.2) to prune these tasks and ask the crowd to do the remaining tasks.

Another important problem is how to efficiently prune the tasks whose difficulties are smaller than a threshold. For crowdsourced joins, a naive way is to enumerate all $\frac{n(n-1)}{2}$ pairs and then compute the difficulty of each pair. There are many indexing techniques such as blocking and Q-gram-based indexing [4] to avoid all-pairs enumeration. Furthermore, when using a similarity-based technique to quantify task difficulties, we can adopt similarity join algorithms [27, 31, 34] to efficiently remove the record pairs whose similarity values are smaller than a threshold.

4.2.3   Pros and Cons

Task pruning is a powerful idea for cost control. It can save human cost by orders of magnitude with only a little loss in quality [25]. Furthermore, it is a very general idea because most crowdsourced operators have already had a lot of computer-only implementations. We can easily design a pruning strategy based on them. On the other hand, this process can be a risky step because if an improper pruning strategy is chosen, then for the tasks that are falsely pruned by computers, they will never be checked by humans again. In addition, it only works for easy tasks (i.e., the ones that machines can do well). In the situation when all the tasks are very challenging, task pruning will not be very effective.

4.3   Answer Deduction

In some cases, the tasks generated by crowdsourced operators have some inherent relationships, which can be leveraged for cost control. Specifically, given a set of tasks, after getting some tasks' results from the crowd, we can use this information to deduce some other tasks' results, saving the cost of asking the crowd to do these tasks. Many crowdsourced operators have such a property, e.g., join [11, 24, 28, 29], planning [14, 35], and mining [2]. For example, suppose a crowdsourced join operator generates three tasks: (A, B), (B, C), and (A, C). If we have already known that A is equal to B and B is equal to C, then we can deduce that A is equal to C based on transitivity, thereby avoiding the crowd cost for checking (A, C).

4.3.1   Iterative Workflow

To leverage answer deduction for cost control, one needs to adopt an iterative workflow to collect answers from the crowd. Figure 4.1 illustrates the workflow. It repeats the following three steps until all tasks' answers are obtained (either through crowdsourcing or deduction).

1. Pick up some tasks without answers from a task pool.
2. Collect their answers from the crowd.
3. Deduce the answers of some other tasks in the pool.

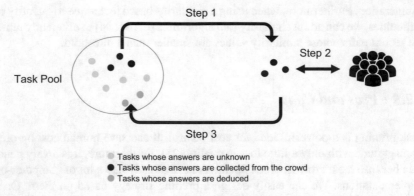

Fig. 4.1 Answer-deduction workflow

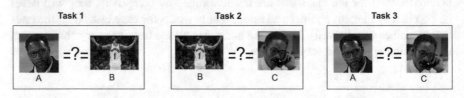

Fig. 4.2 An example for showing the importance of presentation orders

A key challenge to implement this iterative workflow is how to determine in which order tasks should be presented to the crowd. Presentation order often has a big impact on the total cost involved. In the next part, we use crowdsourced joins as an example to illustrate this problem.

4.3.2 Presentation Order

Consider three tasks in Fig. 4.2. Each task is to determine whether two images refer to the same person or not. The ground truth is that $A = B$, $B \neq C$, and $A \neq C$. The following compares the cost of two presentation orders:

- Order 1 (Task 1 → Task 2 → Task 3). Based on the answers of Tasks 1 and 2, we know that $A = B$ and $B \neq C$. Based on transitivity, we obtain the answer of Task 3, i.e., $A \neq C$; thus we do not need to ask the crowd to tell us the answer. In total, this order requires the crowd to do *two* tasks.
- Order 2 (Task 2 → Task 3 → Task 1). Based on the answers of Tasks 2 and 3, we know that $B \neq C$ and $A \neq C$. But they cannot help us deduce the answer of Task 1. In other words, either $A = B$ or $A \neq B$ are possible. In total, this order requires asking the crowd to do *three* tasks.

Therefore, one natural question is what is the optimal presentation order that can maximize the benefit of the use of answer deduction. For crowdsourced joins, it has been proved that the optimal presentation order is to first present matching pairs to the crowd and then present nonmatching pairs [28]. But, this cannot be achieved in practice because it is unknown which pairs are matching or nonmatching upfront. Hence, a heuristic approach [28] is proposed to approximate this ideal case, which first computes a similarity value for each candidate pair and then presents the candidate pairs to the crowd in a decreasing order of similarity values. This approach works very well when there exists a good similarity function to compute the similarity values. To handle the case that such a similarity function does not exist, some other heuristic approaches have been proposed [24], which can provide a better worst-case performance guarantee than the similarity-based one.

4.3.3 Pros and Cons

Answer deduction avoids having the crowd do a lot of redundant work. It even works when only difficult tasks are left. For example, consider the three tasks in Fig. 4.2. They are all very difficult to do, but answer deduction can still save the human cost in this situation. However, answer deduction may hurt both quality and latency. In terms of quality, for example, consider two tasks, where the ground true is $A = B$ and $B = C$. If the crowd falsely labels them as $A = B$ and $B \neq C$, the error will be propagated through transitivity, resulting in an erroneous label of $A \neq C$. Some ideas such as using correlation clustering [29] or designing new decision functions [11] have been explored to tackle this issue. In terms of latency, in order to leverage transitivity, we cannot present all candidate pairs to the crowd at the same time. Instead, it should be an iterative process, where only a single pair or a few pairs can be presented to the crowd at each iteration. As a result, the iterative process may take much longer time to complete because it does not fully utilize all available crowd workers. This is considered as a main challenge in the use of transitivity for crowdsourced ER, and various parallel algorithms are proposed to accelerate the process [28, 29].

4.4 Task Selection

Task selection has been introduced in Sect. 3.3 as an idea to improve quality. From another point of view, this can also be seen as a class of technique to reduce cost. That is, given a quality constraint, the task selection technique can minimize human cost to meet the quality requirement, by selecting the most beneficial tasks for humans to do. Different crowdsourced operators need different task-selection strategies. Due to its effectiveness for cost control, it has been widely studied in a large variety of crowdsourced operators, such as join [10, 13, 18, 23, 30], top-k/sort [3, 7, 12, 15, 20, 33], and categorize [19]. Note that these works are complementary to task assignment introduced in Sect. 3.3 because for a given crowdsourced operator, we first apply a task-selection strategy to decide a set of most beneficial tasks; once these tasks are selected and sent to a crowdsourcing platform, a task-assignment strategy is then used to collect high-quality answers from the crowd.

Most task-selection strategies are *model-driven*. That is, the goal of task selection is to train a machine-learning model (e.g., train a classifier for entity resolution). Thus, they select tasks based on how beneficial these tasks are to the model. In some other situations, the objective may be not to train a model but to solve a specific problem (e.g., find the most beautiful picture in an album). We call this *problem-driven task-selection strategy*, which selects tasks based on how beneficial these tasks can be used to solve a specific problem.

4.4.1 Model-Driven

Model-driven task-selection strategy is referred to as *query strategy* in the active learning literature.

Definition 4.2 (Query Strategy) A query strategy is an algorithm for determining which data points should be labeled.

Various query strategy frameworks have been proposed in the machine learning community [22]. We next present uncertainty sampling, one of the most widely used query strategies. Suppose we want to build a classifier to determine whether an image is of a cat or not. Given N unlabeled images, one solution, called supervised learning, is to label all the images first and then train a model on the labeled data. If N is very large, this approach will be highly expensive. To reduce the cost, we can adopt active learning to select a small number of most beneficial images to train the model. Different query strategies have different ways to decide which images should be selected. In uncertainty sampling, it will select the images that the current model is most uncertain about. Figure 4.3 illustrates this idea. In the figure, the boundary

Fig. 4.3 An illustration of uncertainty sampling

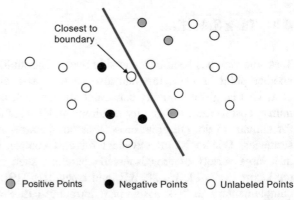

○ Positive Points   ● Negative Points   ○ Unlabeled Points

(i.e., the red line) represents the current model. Uncertainty sampling selects the data point that is most close to the boundary and ask the crowd to label it.

In addition to uncertainty sampling, there are some research works in crowd-sourcing that adopt more sophisticated query strategies for task selection, such as query-by-committee [10] or expected error reduction [18]. Interested readers can refer to Settles [22] for further details about these query strategies.

4.4.2 Problem-Driven

Unlike model-driven, a problem-driven query strategy aims to combine machine outputs and human answers to solve a specific problem. Machines are first applied to get a not-so-good solution to the problem; humans are then used to improve the solution as much as possible.

Consider a specific problem in Fig. 4.4: given five pictures, which one visualizes the best SFU Campus? For this problem, each task is to ask the crowd to compare two pictures and return the more beautiful one. Note that the output of this problem is a single picture rather than a machine-learning model. Therefore, when deciding which task should be selected, the objective is to maximize the chance of returning the most beautiful picture rather than maximize the improvement of a model. Figure 4.4a shows the initial ranking of the five pictures derived from machines. If only one task is allowed, then the most beneficial task should compare the top two pictures (see Fig. 4.4b) because the first two pictures have the highest chance to be the best picture, while others have low possibilities.

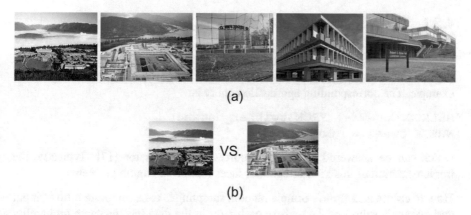

(a)

VS.

(b)

Fig. 4.4 Which picture visualizes the best SFU campus? (**a**) Rank by machines. (**b**) The most beneficial task

4.4.3   Pros and Cons

Task selection provides a flexible way to tune the trade-off between cost and quality. One can easily improve quality by using the technique to select more tasks or reduce the number of selected tasks to save cost. A side effect is that it might increase latency. This is because many task selection techniques need to iteratively query the crowd to decide which tasks can be selected next. In this iterative process, the crowd can only do a small number of tasks per iteration, which ignores the fact that there is a large pool of crowd workers available.

4.5   Sampling

Sampling is a powerful tool to enable fast analytics on large data. Imagine one collects a large twitter dataset and wants to know how many positive tweets are there in the dataset. She can count how many positive tweets in a sample data and then divide the number by the sampling ratio to get an estimate of the true answer. Since only a sample of tweets are processed, the cost will be significantly reduced. This idea inspires some works to leverage sampling for cost control [17, 26]. In these works, the crowd will be first used to process a sample of data, and then statistical estimators will then be applied to estimate final answers from the sample. We next introduce two research topics, crowdsourced aggregation [17] and data cleaning [26], that have adopted this idea.

4.5.1   Crowdsourced Aggregation

Crowdsourced aggregation aims to use the crowd to answer an aggregation query (e.g., count, sum, avg) over a collection of unlabeled items. Consider the above example. The corresponding aggregation query is:

```
SELECT COUNT(*) FROM twitter_dataset
WHERE tweet = "positive";
```

which can be answered by a crowdsourced count operator [17]. Typically, the implementation of this kind of operator faces three challenging problems.

How to create a sample?   Simple random sampling takes a sampling ratio as input and outputs a subset of data where each item in the data has the same probability (equal to the sampling ratio) to be sampled. For example, suppose the sampling ratio is 1%. Then, 1% of the tweets will be uniformly selected. This method is simple but may not be very effective when dealing with a group-by query. For example, suppose one wants to know how many positive tweets are posted from each country, by issuing the query as follows:

```
SELECT COUNT(*) FROM twitter_dataset
WHERE tweet = "positive"
GROUP BY country;
```

If every group uses the same sampling ratio, some small groups may have very few or even no tweets appear in the sample, leading to highly inaccurate estimated counts. Stratified sampling is designed to overcome this limitation. It divides data into multiple groups and applies simple random sampling (often with a different sampling ratio) to each group independently. For example, consider two groups: the USA has 1 million tweets and Sweden has 10,000 tweets. Stratified sampling may apply the sampling ratios of 0.1% and 10% to the two groups, respectively, such that each group has the same number of tweets selected.

How to estimate an answer? Once a sample is created, crowd workers are asked to check each tweet in the sample and determine whether it is positive or not. The final answer will be estimated as:

```
SELECT COUNT(*)/θ FROM twitter_sample_dataset
WHERE tweet = "positive";
```

where θ denotes the sampling ratio. If crowd workers did not make any mistakes, this estimator would be unbiased, i.e., the expected value of the estimated answer is equal to the true answer. Therefore, a key challenge is how to improve worker quality. In addition to the general quality-control techniques presented in Chap. 3, there are a number of other techniques that are particularly tailored to crowdsourced aggregation such as avoiding coordinated attacks [17].

How to quantify uncertainty? Since an answer is estimated based on a sample, one natural question is how far the estimated answer is away from the true answer. A confidence interval is an interval estimation of the true answer. The larger the interval, the more uncertain the estimated answer. For example, a 95% confidence interval of 1000 ± 10 means that the true answer is within the range of 1000 ± 10 with 95% probability. There are two ways to compute a confidence interval. One is to use the central limit theorem to derive its closed form; the other is to employ the bootstrap to compute it empirically. Both have been extensively studied in statistics. Interested readers can refer to [6, 16] for more details.

4.5.2 Data Cleaning

Real-world data is often dirty. Analyses without *data cleaning* can be very risky, which may result in poor decision-making, and have a significant negative impact on applications.

Definition 4.3 (Data Cleaning) Data cleaning is the process of detecting and correcting dirty (inconsistent, inaccurate, or incomplete) values from a database.

Rakesh Agrawal
Microsoft
Publications: 353   Citations: 33537
Fields: Databases, Data Mining, World Wide Web
Collaborated with 365 co-authors from 1982 to 2012 | Cited by 24220 authors

Jeffrey D. Ullman
Stanford University
Publications: 460   Citations: 43431
Fields: Databases, Algorithms & Theory, Scientific Computing
Collaborated with 317 co-authors from 1961 to 2012 | Cited by 31987 authors

Michael Franklin
University of California Berkeley
Publications: 561   Citations: 15174
Fields: Databases, Pharmacology, Data Mining
Collaborated with 3451 co-authors from 1974 to 2012 | Cited by 15795 authors

Fig. 4.5   Who published more?

On the other hand, data cleaning is a costly and time-consuming process. Although there has already been a long line of work on machine-based data cleaning techniques, many cleaning tasks are too challenging for machine-only solutions [8, 21, 32]. The use of crowdsourcing can mitigate this issue but is very expensive to scale to big data. SampleClean [26] is a framework that allows the crowd to only clean a small sample of data and uses the cleaned sample to obtain high-quality results from the full data.

Consider the example in Fig. 4.5. Suppose a data scientist wants to know *who published more among three database researchers* based on Microsoft Academic Search data. If she did not clean any data, the ranking would be Michael Franklin (561) > Jeffrey D. Ullman (460) > Rakesh Agrawal (353). However, the data has many errors. First, there are some papers that are not published by them but are falsely included into their publication lists. Second, there are some duplicate papers in their publication lists. It is hard to infer the impact of these errors on the result. In fact, after manually cleaning all the data, the true ranking should be Jeffrey D. Ullman (255) > Rakesh Agrawal (211) > Michael Franklin (173).

To address this problem, SampleClean finds that there are two sources of errors that may affect a query result: data error and sampling error. If data is not cleaned at all, the query result over the full dirty data will suffer from data error; if a sample of data is cleaned, the query result estimated based on the cleaned sample will suffer from sampling error. Compared to data error, a good property of sampling error is that its value is proportional to $\frac{1}{\sqrt{|S|}}$, where $|S|$ is the sample size. That is, by cleaning a small sample dataset, sampling error can be significantly reduced. For example, if the cleaned sample size is increased from 10 to 1000, the sampling error can be decreased by 10 times. More importantly, sampling error can be bounded by a confidence interval. Being able to know how uncertain about estimated answers is essential to reliable data-driven decision-making.

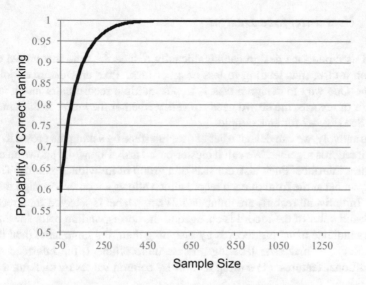

Fig. 4.6 The effectiveness of SampleClean for answering the question in Fig. 4.5

Figure 4.6 shows the performance of SampleClean on the Microsoft Academic Search dataset. We can see that SampleClean can have a probability of 95% to get the correct ranking by only cleaning about 200 records. Note that without any data cleaning, we will always get the wrong ranking. After cleaning 400 records, SampleClean almost always returns the correct ranking, which is much cheaper than cleaning the full data (1374 records).

4.5.3 Pros and Cons

Sampling is a very powerful tool for cost control. Decades of research on sample estimates has built good theories that can effectively bound the statistical error of the estimates. However, a drawback is that sampling does not work for all crowdsourced operators. For example, when applying sampling to a crowdsourced max operator, the estimated max from the sample may deviate a lot from the true max value.

4.6 Task Design

Unlike the previous cost-control techniques, task design seeks to reduce c (i.e., the monetary cost to complete one task) rather than n (i.e., the number of tasks). This section introduces two ideas to achieve this goal. The first one is through better user interface design; the other one is to incentivize the crowd in a non-monetary way (e.g., fun, community participation).

4.6.1   User Interface Design

A good user interface design can significantly reduce the time that crowd workers spend on a task, thus leading to less cost per task. Take crowdsourced join as an example. One way to design a task is to put multiple record pairs into a task. For each pair of records, the crowd needs to verify whether they refer to the same entity or not. See Fig. 4.7 for an example.

Alternatively, we can design a better user interface by asking the crowd to verify a group of records together. We call it cluster-based task. Figure 4.8 shows an example of the user interface. Each task consists of a group of individual records. There is a drop-down list at the front of each record which allows a worker to assign the record a label. Initially, all records are unlabeled. When a label is selected for a record, the background color of the record is changed to the corresponding color for that label. Workers indicate matching records by assigning them the same label (and thus, the same color). To make the labeling process more efficient, the interface supports two additional features: (1) sorting records by column values by clicking a column

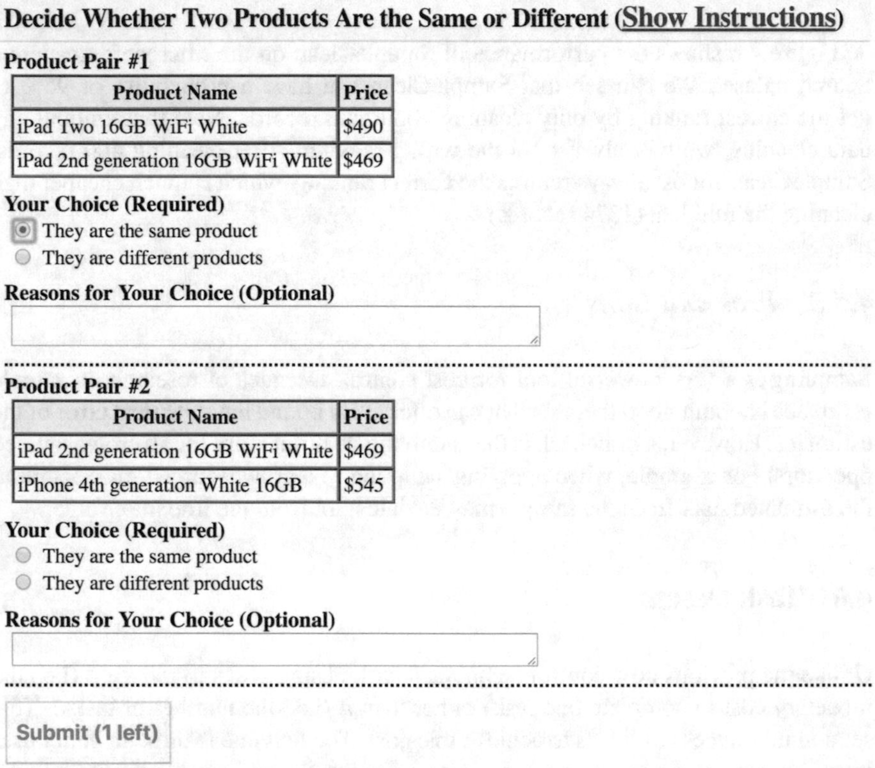

Decide Whether Two Products Are the Same or Different (Show Instructions)

Product Pair #1

Product Name	Price
iPad Two 16GB WiFi White	$490
iPad 2nd generation 16GB WiFi White	$469

Your Choice (Required)
- ⦿ They are the same product
- ○ They are different products

Reasons for Your Choice (Optional)

· ·

Product Pair #2

Product Name	Price
iPad 2nd generation 16GB WiFi White	$469
iPhone 4th generation White 16GB	$545

Your Choice (Required)
- ○ They are the same product
- ○ They are different products

Reasons for Your Choice (Optional)

· ·

Submit (1 left)

Fig. 4.7 A pair-based task with two pairs of records

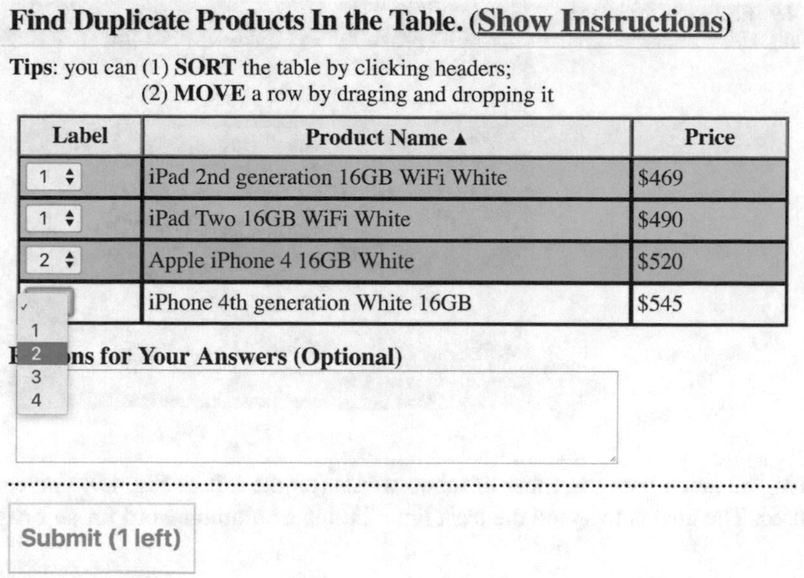

Find Duplicate Products In the Table. (Show Instructions)

Tips: you can (1) **SORT** the table by clicking headers;
(2) **MOVE** a row by draging and dropping it

Label	Product Name ▲	Price
1 ⬍	iPad 2nd generation 16GB WiFi White	$469
1 ⬍	iPad Two 16GB WiFi White	$490
2 ⬍	Apple iPhone 4 16GB White	$520
✓	iPhone 4th generation White 16GB	$545

1
2
3
4

ns for Your Answers (Optional)

Submit (1 left)

Fig. 4.8 A cluster-based task with four records

header and (2) moving a record by dragging and dropping it. The first feature can be used, for example, to sort the records based on a specific attribute such as product price. The second feature can be used, for example, to place the records that share a common word, e.g., "iPad," near each other for easier comparison. We can see that by doing a cluster-based task, it is equivalent to verifying n^2 pairs but with much less effort.

4.6.2 Non-monetary Incentives

Many people choose to do some jobs not only for money but also for fun, learning, reputation, etc. If we can find a non-monetary way to incentivize the crowd, the monetary cost can be substantially reduced.

For example, ESP game [1] is a crowdsourcing game developed for image labeling. Given an image that needs to be labeled, the game will randomly match two players and ask them to use some words to describe the image. Note that the two players do not know each other and they cannot communicate. Figure 4.9 shows an example of the game's user interface that one player interacts with. The player has guessed two words: "tree" and "grass." If the other player has also used either one of them (e.g., "tree") to describe the image, they win the game and the common word (i.e., "tree") will be selected as a label for the image. To avoid them selecting easy

Fig. 4.9 ESP game for image labeling

words, the game lists a number of taboo words (on the left of Fig. 4.9) that cannot be used. The goal is to spend the least time finding a common word for an image.

4.6.3 Pros and Cons

Compared to the other cost-control techniques presented in this chapter, task design is focused on a different angle to optimize crowd cost. This unique feature enables it to integrate with other techniques seamlessly and to help a requester further reduce cost. However, it also has two limitations. First, the crowd needs to spend additional time in learning an unfamiliar interface. Many workers may not want to do so [25] although the new interface can save their time. Second, it is hard to keep workers feeling excited about one thing for a long period of time; thus the number of interested workers may decrease dramatically as time goes on. For example, most people may not be willing to keep playing the same game for a long time.

4.7 Summary of Cost Control

Some cost-control techniques are used before a requester publishes tasks, e.g., task pruning and sampling, while some techniques can be used iteratively, e.g., answer deduction and task selection. In fact, these cost-control techniques can be used together. For example, task pruning can be first used to prune a lot of easy tasks. Then, for the remaining tasks, task selection can be utilized to decide which tasks should be selected for the crowd.

In addition, there is a trade-off between quality and cost. The cost-control techniques may sacrifice quality. For example, answer deduction may reduce quality if the crowd makes some mistakes in their answers, and task pruning can decrease quality if some important tasks are pruned as discussed above. Thus, when using a cost-control technique, it is important to consider how to balance the trade-off between quality and cost [3, 25, 28].

References

1. von Ahn, L., Dabbish, L.: ESP: labeling images with a computer game. In: AAAI, pp. 91–98 (2005)
2. Amsterdamer, Y., Davidson, S.B., Milo, T., Novgorodov, S., Somech, A.: Oassis: query driven crowd mining. In: SIGMOD, pp. 589–600. ACM (2014)
3. Chen, X., Bennett, P.N., Collins-Thompson, K., Horvitz, E.: Pairwise ranking aggregation in a crowdsourced setting. In: WSDM, pp. 193–202 (2013)
4. Christen, P.: A survey of indexing techniques for scalable record linkage and deduplication. IEEE Trans. Knowl. Data Eng. 24(9), 1537–1555 (2012)
5. Deng, D., Li, G., Feng, J.: A pivotal prefix based filtering algorithm for string similarity search. In: SIGMOD, pp. 673–684 (2014)
6. Efron, B., Tibshirani, R.J.: An introduction to the bootstrap. CRC press (1994)
7. Eriksson, B.: Learning to top-k search using pairwise comparisons. In: AISTATS, pp. 265–273 (2013)
8. Fan, W., Li, J., Ma, S., Tang, N., Yu, W.: Towards certain fixes with editing rules and master data. PVLDB 3(1), 173–184 (2010)
9. Feng, J., Wang, J., Li, G.: Trie-join: a trie-based method for efficient string similarity joins. VLDB J. 21(4), 437–461 (2012)
10. Gokhale, C., Das, S., Doan, A., Naughton, J.F., Rampalli, N., Shavlik, J.W., Zhu, X.: Corleone: hands-off crowdsourcing for entity matching. In: SIGMOD, pp. 601–612 (2014)
11. Gruenheid, A., Kossmann, D., Ramesh, S., Widmer, F.: Crowdsourcing entity resolution: When is A=B? Technical report, ETH Zürich
12. Guo, S., Parameswaran, A.G., Garcia-Molina, H.: So who won?: dynamic max discovery with the crowd. In: SIGMOD, pp. 385–396 (2012)
13. Jeffery, S.R., Franklin, M.J., Halevy, A.Y.: Pay-as-you-go user feedback for dataspace systems. In: SIGMOD, pp. 847–860 (2008)
14. Kaplan, H., Lotosh, I., Milo, T., Novgorodov, S.: Answering planning queries with the crowd. PVLDB 6(9), 697–708 (2013)
15. Khan, A.R., Garcia-Molina, H.: Hybrid strategies for finding the max with the crowd. Tech. rep. (2014)
16. Lohr, S.: Sampling: design and analysis. Nelson Education (2009)
17. Marcus, A., Karger, D.R., Madden, S., Miller, R., Oh, S.: Counting with the crowd. PVLDB 6(2), 109–120 (2012)
18. Mozafari, B., Sarkar, P., Franklin, M., Jordan, M., Madden, S.: Scaling up crowd-sourcing to very large datasets: a case for active learning. PVLDB 8(2), 125–136 (2014)
19. Parameswaran, A.G., Sarma, A.D., Garcia-Molina, H., Polyzotis, N., Widom, J.: Human-assisted graph search: it's okay to ask questions. PVLDB 4(5), 267–278 (2011)
20. Pfeiffer, T., Gao, X.A., Chen, Y., Mao, A., Rand, D.G.: Adaptive polling for information aggregation. In: AAAI (2012)
21. Sarawagi, S., Bhamidipaty, A.: Interactive deduplication using active learning. In: SIGKDD, pp. 269–278 (2002)
22. Settles, B.: Active learning literature survey. University of Wisconsin, Madison 52(55–66), 11
23. Verroios, V., Garcia-Molina, H.: Entity resolution with crowd errors. In: ICDE, pp. 219–230 (2015)

24. Vesdapunt, N., Bellare, K., Dalvi, N.N.: Crowdsourcing algorithms for entity resolution. PVLDB **7**(12), 1071–1082 (2014)
25. Wang, J., Kraska, T., Franklin, M.J., Feng, J.: CrowdER: crowdsourcing entity resolution. PVLDB **5**(11), 1483–1494 (2012)
26. Wang, J., Krishnan, S., Franklin, M.J., Goldberg, K., Kraska, T., Milo, T.: A sample-and-clean framework for fast and accurate query processing on dirty data. In: SIGMOD, pp. 469–480 (2014)
27. Wang, J., Li, G., Feng, J.: Can we beat the prefix filtering?: an adaptive framework for similarity join and search. In: SIGMOD, pp. 85–96 (2012)
28. Wang, J., Li, G., Kraska, T., Franklin, M.J., Feng, J.: Leveraging transitive relations for crowdsourced joins. In: SIGMOD, pp. 229–240 (2013)
29. Wang, S., Xiao, X., Lee, C.: Crowd-based deduplication: An adaptive approach. In: SIGMOD, pp. 1263–1277 (2015)
30. Whang, S.E., Lofgren, P., Garcia-Molina, H.: Question selection for crowd entity resolution. PVLDB **6**(6), 349–360 (2013)
31. Xiao, C., Wang, W., Lin, X., Yu, J.X., Wang, G.: Efficient similarity joins for near-duplicate detection. ACM Trans. Database Syst. **36**(3), 15:1–15:41 (2011)
32. Yakout, M., Elmagarmid, A.K., Neville, J., Ouzzani, M., Ilyas, I.F.: Guided data repair. PVLDB **4**(5), 279–289 (2011)
33. Ye, P., EDU, U., Doermann, D.: Combining preference and absolute judgements in a crowd-sourced setting. In: ICML Workshop (2013)
34. Yu, M., Li, G., Deng, D., Feng, J.: String similarity search and join: a survey. Frontiers of Computer Science **10**(3), 399–417 (2016)
35. Zhang, C.J., Tong, Y., Chen, L.: Where to: Crowd-aided path selection. PVLDB **7**(14), 2005–2016 (2014)

Chapter 5
Latency Control

Latency refers to the total time of completing a job. Since humans are slower than machines, sometimes even a simple job (e.g., labeling one thousand images) may take hours or even days to complete. Thus, another big challenge in crowdsourced data management is latency control, that is, how to reduce job completion time while still keeping good result quality as well as low cost.

This chapter first classifies existing latency-control techniques into three categories (single-task latency control, single-batch latency control, and multi-batch latency control) and then presents an overview of these techniques in Sect. 5.1. After that, it describes how to reduce the latency of a single task in Sect. 5.2, the latency of a batch of tasks in Sect. 5.3, and the latency of multiple batches of tasks in Sect. 5.4, respectively.

5.1 Overview of Latency Control

Inspired by cost control, one may be tempted to calculate job completion time as latency $= n \cdot t$, where n is the number of tasks and t is the average time spent on each task. This formula, however, is not correct because it neglects the fact that crowd workers do tasks in parallel. The correct formula should be

$$\text{latency} = t_{\text{last}},$$

where t_{last} represents the completion time of the last task. For example, suppose there are 10 workers and their job is to label 1,000 images. The latency of this job is the time when the last image has been labeled. This is similar to parallel computing where the time is determined by the slowest node.

We classify existing latency-control techniques into three categories. Figure 5.1 illustrates their differences.

© Springer Nature Singapore Pte Ltd. 2018
G. Li et al., *Crowdsourced Data Management*,
https://doi.org/10.1007/978-981-10-7847-7_5

Fig. 5.1 Three classes of existing latency-control techniques

① *Single-task latency control* aims to reduce the latency of one task (e.g., the latency of labeling each individual image).

② *Single-batch latency control* aims to reduce the latency of a batch of tasks (e.g., the latency of labeling ten images at the same time).

③ *Multi-batch latency control* aims to reduce the latency of multiple batches of tasks (e.g., adopting an iterative workflow to label a group of images where each iteration labels a batch of two images).

Note that since ① is a special case of ② and ② is a special case of ③, all the techniques proposed for ① can be used for ②, and all the techniques proposed for ② can be used for ③.

5.2   Single-Task Latency Control

Assuming a requester only has a single task, how can she get the task's answer as soon as possible? In this situation, task latency consists of three parts: recruitment time, qualification test time, and work time.

5.2.1   Recruitment Time

This section first considers the recruitment time.

Definition 5.1 (Recruitment Time) Recruitment time refers to the time from a task being published (by a requester) until it being accepted (by a worker).

For example, consider the task of labeling an image. Once the task is published, it will not be accepted by workers immediately. This is because in a crowdsourcing platform, there are a large number of other tasks available. Workers will consider many factors such as task price, task complexity, or requester reputation and then decide which task to do. Suppose it takes 10 min for the task to be accepted, then the recruitment time is 10 min.

To reduce the recruitment time, Bernstein et al. [1] proposed a technique called *retainer model*. The basic idea is to spend a small amount of money maintaining a worker pool. Once a task is available, the workers in the pool will get notified, and can respond to the task quickly. While the idea sounds simple, the challenge is how to implement it on a crowdsourcing platform (e.g., AMT). To maintain a pool

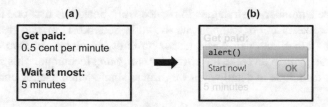

Fig. 5.2 An illustration of the retainer model. (**a**) Workers sign up in advance. (**b**) Alert when task is ready

of workers, the retainer model creates a special task like the one in Fig. 5.2a, which tells workers that they can get 0.5 cent per minute by staying with the task. Note that workers just need to keep the task page open for getting the money. When a requester publishes an actual task (e.g., labeling an image), an alert will be sent to all the workers in the pool notifying that there is a new task available, and click "OK" to view it (Fig. 5.2b). Experimental results show that the retainer model can reduce recruitment time from minutes to less than 2 seconds, enabling real-time crowdsourcing [1].

5.2.2 Qualification Test Time

The purpose of a qualification test is to prevent malicious workers. It contains a set of predefined tasks whose ground truths are known. A worker has to complete these tasks and get a score (e.g., the percentage of correctly answered tasks). If the score is lower than a requester's requirement, the worker fails the test and will not be allowed to do other tasks.

The use of a qualification test may have a negative impact on task latency for two reasons. Firstly, many crowd workers do not like to accept a task that requires passing a qualification test. Secondly, workers need to spend additional time working on a qualification test, increasing overall work time.

To overcome these limitations, the qualification tests can be replaced with other quality-control techniques. For example, a task can be assigned to multiple workers and then use a truth inference approach to determine the final answer. Alternatively, a task assignment approach can be used to detect high-quality workers and then only assign tasks to these workers. More details about these techniques can be found in Chap. 3.

5.2.3 Work Time

Work time is the actual time that workers spend on doing a task.

Definition 5.2 (Work Time) Work time refers to the time from a task being accepted to it being completed.

There are a number of strategies to reduce work time. The first one is to design a better user interface. For example, as shown in Sect. 4.6.1, we can create cluster-based tasks (rather than pair-based tasks) for entity resolution to save work time. Another strategy is to leverage the *straggler mitigation* technique. This section will present this technique in detail when we discuss single-batch latency control.

5.3 Single-Batch Latency Control

The goal of single-batch latency control is to reduce the completion time of the last task in a batch. We group existing works into two categories. *Statistical models* seek to model task latency using statistics and then adopt different strategies (e.g., pricing) to improve task latency. *Straggler mitigation* does not need to predict task latency. It leverages the idea of task redundancy to mitigate the impact of slow tasks. In comparison, straggler mitigation tends to be more robust in practice.

5.3.1 Statistical Model

Some existing works [2, 7] collect statistics from real-world crowdsourcing plat-forms and use such information to model workers' behaviors. Yan et al. [7] build statistical models to predict the time of answering a task. They consider two delays: the delay for the arrival of the first response and the inter-arrival time between two responses. (1) The first delay captures the interval starting from the time of posting a task to the time of receiving the first answer. (2) The second delay, i.e., inter-arrival time, models the interval from receiving adjacent answers of a task. Faradani et al. [2] use statistical models to predict worker's arrival rate in a crowdsourcing platform and characterize how workers select tasks from the platform.

With the assistance of the statistical models, some works [3, 7] study latency control by considering whether or not a task can be accomplished within a time constraint and evaluating the benefit of publishing a new task with respect to particular goals. For example, Gao and Parameswaran [3] study how to set task price in order to meet latency or cost constraints. They propose novel pricing algorithms based on decision theory and show that a dynamic pricing technique can be much more cost-saving than simply setting each task with a fixed price.

5.3.2 Straggler Mitigation

A major drawback of the use of statistical models is that they cannot handle stragglers very well.

Definition 5.3 (Straggler) Stragglers refer to the tasks that are much slower than the others.

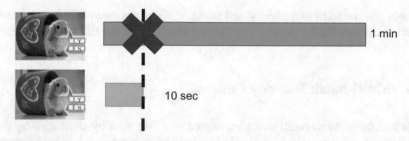

Fig. 5.3 An illustration of straggler mitigation

Predicting stragglers based on workers' historical performance can be problematic. For example, consider a worker who often does tasks very quickly. A task assigned to this worker is very unlikely to be predicated as a straggler. But imagine that the worker suddenly gets a phone call when she is working on a task. It might take her a while to answer the phone and then come back to complete the task. In this situation, the task that is assigned to the worker becomes a straggler. This kind of situation is hard to predict.

Can we mitigate the impact of slow workers without predicting who are the slow workers? The basic idea of straggler mitigation is illustrated in Fig. 5.3. Suppose that we want to label an image and there are two workers available. Rather than predict which one is faster, we ask both of them to label the image. Whenever one of them finishes labeling, we stop the labeling process. In this example, after waiting for 10 s, we get the label from the second worker. We will use it as the final answer without waiting for the response of the other worker.

We now present how to extend this idea to a batch of tasks [4]. Given a pool of workers and a batch of tasks, we call a worker active if she is currently working on a task and available otherwise; similarly, a task is either active, complete, or unassigned.

At the beginning, whenever there is an available worker in the pool, we will route an unassigned task to her. Once all tasks are active or complete, available workers are immediately assigned active tasks, creating duplicate assignments of those tasks. We return the first completed assignment of a task to the requester and immediately reassign all other workers still working on that task to a new active task (though we pay them for their partial work on the old task regardless). The effect of straggler mitigation is that when an inconsistent worker takes a long time to complete a task, the system hides that latency by sending the task to other, faster workers. As a result, the fastest workers complete the majority of the tasks and earn money commensurate with their speed.

A natural question arises when performing straggler mitigation: which task should be assigned to an available worker? Haas et al.[4] ran simulation experiments testing several straggler routing algorithms, including routing to the longest-running active task, to a random task, to the task with fewest active workers, or to the task known by an oracle to complete the slowest. Surprisingly, the selection algorithm did not affect end-to-end latency and is randomly performed as fast as the oracle solution because the fast workers completed almost all of the tasks in the batch.

5.4 Multi-batch Latency Control

This section presents multi-batch latency control. We start by showing why there is a need to publish tasks in multiple batches and then present two basic ideas to control multi-batch latency.

5.4.1 Motivation of Multiple Batches

The main reason to publish tasks in multiple batches is to save cost. In Chap. 4, we introduce various cost-control techniques, where many of them require an iterative workflow. For example, the answer-deduction technique needs to publish tasks iteratively in order to use the answers of already published tasks to deduce the answers of unpublished tasks; the task-selection technique such as active learning needs to iteratively query an oracle to train a good model.

The fundamental problem of multi-batch latency control is how to balance the trade-off between cost and latency. One extreme case is to publish all tasks in a single batch. This is optimal in terms of latency, but misses the opportunity of leveraging an iterative workflow to save cost. The other end of the spectrum is to publish a single task per batch. This is optimal in terms of cost but neglects the fact that the crowd can do tasks in parallel, thus suffering from high latency.

5.4.2 Two Basic Ideas

We now present two basic ideas used by existing work to balance the cost-latency trade-off.

Idea 1. Increase Batch Size Without Hurting Cost The optimal cost can be achieved by publishing a single task per batch. That is, the batch size is one. In some situations, we can increase batch size such that the corresponding cost stays the same. Consider the three entity resolution tasks in Fig. 4.2. Suppose the labeling order is $\langle (A, B), (B, C), (A, C) \rangle$. The naive labeling approach will label one pair at a time. It first asks the crowd to label the first pair (A, B) and cannot crowdsource the second pair until the first pair is labeled. However, for the second pair (B, C), we observe that no matter which label the first pair gets, we must crowdsource it since the second pair cannot be deduced from the first pair (A, B) based on transitivity. For the third pair (A, C), it cannot be published with the first two pairs since if

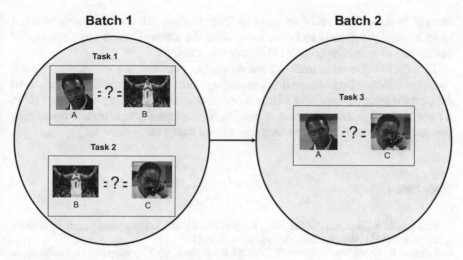

Fig. 5.4 An example of multi-batch latency control

$A = B$ and $B = C$, then its label will be deducted based on transitivity. Thus, as shown in Fig. 5.4, we put the first two pairs to Batch 1 and the third pair to Batch 2. Compared to publishing one task at a time, this strategy can publish multiple tasks at each iteration while still achieving the optimal cost.

Idea 2. Round Model Some existing works [5, 6] use the round model to do latency control. Supposing there are enough active workers, Sarma et al. [5] simplify the definition of latency by assuming that each round spends one unit time, and then the latency is modeled as the number of rounds. Suppose there are n tasks and each round selects k tasks, then it requires n/k rounds in total to finish all tasks. To improve latency, they use the answer-deduction idea in Sect. 4.3 to reduce the number of published tasks, that is, tasks may have relationships, and the answers of tasks received in previous rounds can be leveraged to decide the chosen tasks in the next round. This in fact decreases the total number of tasks to be asked (i.e., $< n$), and if each round selects k tasks that cannot be deduced, the total number of rounds is $< n/k$, thus reducing the latency. Vasilis et al. [6] formalize the problem as a latency budget allocation problem and propose effective optimization techniques to solve the problem.

5.5 Summary of Latency Control

Most existing works in latency control (e.g., straggler mitigation, retainer model, multi-batch latency control) investigate how to balance the trade-off between latency and cost. The similar trade-off also exists between latency and quality. For example, to increase quality, task assignment – a quality-control technique – assigns hard tasks to more workers and easy tasks to fewer workers. To do so, it needs to select tasks in multiple rounds to better understand the tasks. Thus, a large number of rounds can improve the quality but reduce the latency.

In order to balance the trade-off among quality, cost, and latency, existing studies focus on different problem settings, including optimizing the quality given a fixed cost, minimizing the cost with a little sacrifice of quality, reducing the latency given a fixed cost, minimizing the cost within latency and quality constraints, optimizing the quality without taking too much latency and cost, etc.

References

1. Bernstein, M.S., Brandt, J., Miller, R.C., Karger, D.R.: Crowds in two seconds: enabling realtime crowd-powered interfaces. In: UIST, pp. 33–42 (2011)
2. Faradani, S., Hartmann, B., Ipeirotis, P.G.: What's the right price? pricing tasks for finishing on time. In: AAAI Workshop (2011)
3. Gao, Y., Parameswaran, A.G.: Finish them! Pricing algorithms for human computation. PVLDB **7**(14), 1965–1976 (2014)
4. Haas, D., Wang, J., Wu, E., Franklin, M.J.: Clamshell: Speeding up crowds for low-latency data labeling. PVLDB **9**(4), 372–383 (2015)
5. Sarma, A.D., Parameswaran, A.G., Garcia-Molina, H., Halevy, A.Y.: Crowd-powered find algorithms. In: ICDE, pp. 964–975 (2014)
6. Verroios, V., Lofgren, P., Garcia-Molina, H.: tdp: An optimal-latency budget allocation strategy for crowdsourced MAXIMUM operations. In: SIGMOD, pp. 1047–1062 (2015)
7. Yan, T., Kumar, V., Ganesan, D.: Crowdsearch: exploiting crowds for accurate real-time image search on mobile phones. In: MobiSys, pp. 77–90 (2010)

Chapter 6
Crowdsourcing Database Systems and Optimization

It is rather inconvenient to interact with crowdsourcing platforms, as the platforms require one to set various parameters and even write code. Thus, inspired by traditional DBMS, crowdsourcing database systems have been designed and built. Some primary systems developed recently include CrowdDB [6], Qurk [11], Deco [14], CDAS [4], and CDB [8]. The design principle of a crowdsourcing database system is to integrate crowdsourcing into relational database management systems (RDBMS). The system provides declarative programming interfaces and allows requesters to use an SQL-like language for posing queries that involve crowdsourced operations. It leverages crowd-powered operations (aka operators) to encapsulate the complexities of interacting with the crowd. Under this principle, given an SQL-like query from a requester, the system first parses the query into a query plan with crowd-powered operators, then generates tasks to be published on crowdsourcing platforms, and finally collects the crowd's inputs for producing the result.

This chapter first presents an overview of the crowdsourcing database systems in Sect. 6.1. Then, crowdsourcing query language and query optimization techniques are, respectively, introduced in Sects. 6.2 and 6.3. Finally, this chapter is summarized by providing a comprehensive comparison of the existing crowdsourcing database systems in Sect. 6.4.

6.1 Overview of Crowdsourcing Database Systems

Crowdsourcing database systems have two main differences from traditional RDBMS. Firstly, traditional RDBMS use the "close-world" model, processing queries based on data inside the database only, while crowdsourcing database systems use the "open-world" model, utilizing the crowd to crowdsource data, i.e., collecting a tuple/table or filling an attribute. Secondly, crowdsourcing database

© Springer Nature Singapore Pte Ltd. 2018
G. Li et al., *Crowdsourced Data Management*,
https://doi.org/10.1007/978-981-10-7847-7_6

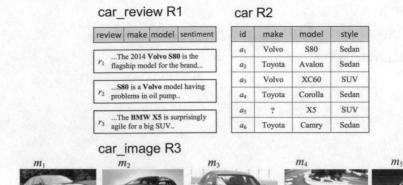

Fig. 6.1 A running example for crowdsourcing database systems

systems can utilize the crowd to answer machine-hard queries, e.g., comparing two objects, ranking multiple objects, and rating an object. For better illustration, we take the database shown in Fig. 6.1 as an example, where relational tables `car`, `car_review` and `car_image`, respectively, contain records of car information, car reviews, and car images. On top of this database, let us consider two queries Q_1 and Q_2 below.

Q_1:   SELECT * FROM `car` R2 WHERE `make` = "BMW"
Q_2:   SELECT * FROM `car` R2, `car_image` R3
     WHERE R2.`make` = R3.`make` AND R2.`model` = R3.`model` AND R2.`color` = "red"

Due to the close-world assumption, traditional RDBMS cannot retrieve any results for both queries. On the contrary, a crowdsourcing database system can utilize the crowd to fill the missing value of tuple or to make record-image comparisons, leading to better query evaluation performance.

As mentioned previously, several crowdsourcing database systems [4, 6, 8, 11, 14] have been recently developed. Next, we present an overview of the design of these systems.

Data Model Existing crowdsourcing database systems are built on top of the traditional *relational* data model, which is defined as follows:

Definition 6.1 (Data Model) The data model of crowdsourcing database systems is the relational model, where data is specified as a schema that consists of relations and each relation has a set of attributes.

However, the difference is that crowdsourcing database systems employ an *open-world* assumption that either some attributes of a tuple or even an entire tuple can be crowdsourced on demand based on queries from the requester.

Crowdsourcing Query Language Most of the crowdsourcing query languages follow the standard SQL syntax and semantics and extend SQL by adding features that support crowdsourced operations.

- *Data define language (DDL)* is used to define how a system invokes crowdsourcing for query processing. The existing systems have designed their own mechanisms. CrowdDB [6] introduces a keyword **CROWD** in a **CREATE TABLE** clause to define which attributes or relation tuples can be crowdsourced. Deco [14] defines a conceptual schema that partitions attributes into *anchor attributes* and *dependent attribute groups* and specifies fetch and resolution rules. Qurk [11] employs user-defined functions (UDFs) to define crowd-based expressions, which can be easily integrated with SQL. CDAS [4] introduces keyword **ByPass** to define which attributes are not needed for crowdsourcing. CDB [8] applies a similar DDL to CrowdDB by also using keyword **CROWD**.
- *Query semantics* is used to express crowdsourcing requirements. Typically, existing systems support the following two types of query semantics.

  Definition 6.2 (Crowd-Powered Collection) Crowd-powered collection solicits the crowd to fill missing attributes of existing tuples or collect more tuples.

  Definition 6.3 (Crowd-Powered Query) Crowd-powered query asks the crowd to perform data processing operations, such as selection, join, sort, etc. on the underlying data.

  Moreover, due to the open-world nature of crowdsourcing, some systems [4, 14] also define constraints on the number of returned results or the cost budget for crowdsourcing.

The details of crowdsourcing query languages will be presented in Sect. 6.2.

Crowdsourcing Query Processing The architecture of a typical crowdsourcing database system is illustrated in Fig. 6.2. An SQL-like query is issued by a crowdsourcing requester and is first processed by a query optimizer.

Definition 6.4 (Crowdsourcing Optimizer) A crowdsourcing optimizer parses a query into a structured query plan and then applies optimization strategies to produce an *optimized* query plan.

However, the key difference is that the tree nodes in a query plan are *crowd-powered operators*. Typically, a crowd-powered operator abstracts a specific type of operation that can be processed by the crowd. Recent years have witnessed many studies on developing crowd-powered operators, such as crowd-powered selection (**CrowdSelect**) [12, 17, 21], join (**CrowdJoin**) [19, 20], collect (**CrowdCollect** and **CrowdFill**) [16, 18], top-k/sort (**CrowdSort** and **CrowdTopK**) [3, 10], and aggregation (**CrowdCount**, **CrowdMax**, and **CrowdMin**) [7, 9]. Implementation of these operators will be introduced in Chap. 7.

Then, crowd-powered operators are executed by the **crowdsourcing executor**.

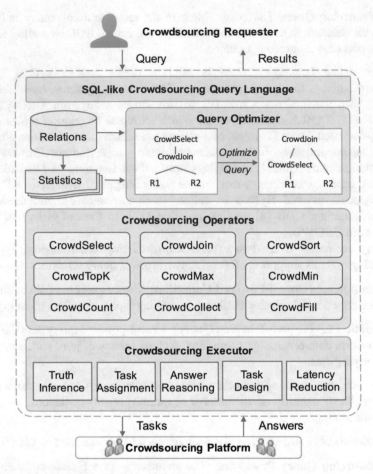

Fig. 6.2 Architecture of crowdsourcing DB systems

Definition 6.5 (Crowdsourcing Executor) A crowdsourcing executor generates Human Intelligent Tasks (HITs) and publishes the HITs on crowdsourcing platforms (e.g., AMT). After collecting answers from the crowd, it evaluates the query plan and returns the final results.

To this end, the executor employs several crowdsourcing data processing techniques, i.e., truth inference, task assignment, answer reasoning, task design, and latency reduction, which were discussed in previous chapters.

Crowdsourcing Query Optimization Query optimization is indispensable in crowdsourcing database systems, as the difference of various query plans may be several orders of magnitude. It is worth noting that crowdsourcing optimization is more challenging than that of traditional databases because it needs to optimize multiple crowdsourcing objectives, (1) quality (better results), (2) cost (less money), and (3) latency (higher speed). It is desirable for a system to support "multi-objective" optimization, as any single optimization may not satisfy a requester's needs. Crowdsourcing query optimization techniques will be introduced in Sect. 6.3.

6.2 Crowdsourcing Query Language

The basic data model in crowdsourcing database systems is the relational model. On top of this, existing systems develop *crowdsourcing query languages*.

Definition 6.6 (Crowdsourcing Query Language) A crowdsourcing query language is designed by extending classic SQL with new features to support crowd-powered operations, so as to encapsulate the complexities of interacting with the crowd.

One advantage of this design is that SQL programmers can easily learn to use crowdsourcing database systems, while user-friendly SQL query formulation tools can be easily applied to inexperienced users. In this section, we highlight the differences between the mechanisms of designing the crowdsourcing languages of CrowdDB [6], Qurk [11], Deco [14], CDAS [4], and CDB [8].

6.2.1 CrowdDB

The CrowdDB system [5, 6] designs a query language called CrowdSQL to support crowdsourcing missing data and performing crowd-powered comparisons. CrowdSQL is a natural extension of standard SQL with three newly introduced keywords **CROWD**, **CROWDEQUAL**, and **CROWDORDER**.

- Keyword **CROWD** is introduced in the DDL of CrowdSQL for defining which part of data can be crowdsourced, so as to allow CrowdDB to collect missing data on demand from the crowd. As illustrated in Fig. 6.3a, missing data can occur in two cases: (1) specific attribute values of tuples could be crowdsourced, and (2) entire tuples could be crowdsourced. CrowdSQL uses **CROWD** keywords to specify both cases in the **CREATE TABLE** statement of CrowdSQL. Note that **CROWD** can be used to specify any column or table in CrowdDB and it can also be integrated with other features of **CREATE TABLE** syntax, e.g., integrity constraints.
- Keywords **CROWDEQUAL** and **CROWDORDER** are introduced in the query semantics of CrowdSQL for involving the crowd to perform machine-hard comparisons. For example, the query in the left part of Fig. 6.3b utilizes

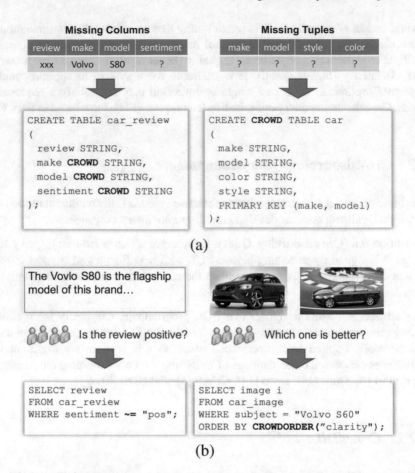

Fig. 6.3 CrowdSQL: Crowdsourcing language of CrowdDB. (**a**) **CROWD** for crowdsourcing missing data. (**b**) **CROWDEQUAL** and **CROWDORDER** in query semantics

CROWDEQUAL (represented as symbol $\sim=$) to identify car reviews that satisfy certain constraints, i.e., sentiment being "positive" by soliciting the crowd. Another crowd-powered operation is **CROWDORDER**, which is illustrated in the right part of Fig. 6.3b. Given a set of tuples which are comparable but are hard to be compared by machines, **CROWDORDER** is used to crowdsource the ordering tasks based on certain criterion, e.g., "clarity" in the query.

6.2.2   Qurk

The Qurk system [11] also extends SQL to encapsulate the input from the crowd, in order to make its query language familiar to a database audience. However,

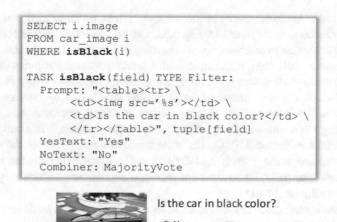

```
SELECT i.image
FROM car_image i
WHERE isBlack(i)

TASK isBlack(field) TYPE Filter:
  Prompt: "<table><tr> \
      <td><img src='%s'></td> \
      <td>Is the car in black color?</td> \
      </tr></table>", tuple[field]
  YesText: "Yes"
  NoText: "No"
  Combiner: MajorityVote
```

Fig. 6.4 Crowdsourcing query language of Qurk

it introduces a different way, i.e., utilizing user-defined functions (UDFs). A UDF in Qurk can be considered as a *template* to be instantiated for generating crowdsourcing tasks. As shown in Fig. 6.4, the query defines a UDF **isBlack** for asking the crowd to identify black cars.

Typically, a UDF in Qurk contains the following fields:

- The **TYPE** field pre-defines some kinds of crowdsourcing tasks, including (1) *filter* that selects tuples satisfying some filtering conditions, (2) *sort* that ranks tuples according to some criterion defined in the UDF, (3) *join* that connects tuples from multiple tables, and (4) *generative* that collects data from the crowd. For example, the UDF in Fig. 6.4 defines crowdsourcing tasks with type *filter*.
- The parameters for task UI generation, e.g., **Prompt**, **YesText**, and **NoText**, specify parameterized HTML texts for generating task user interfaces.
- The **combiner** invokes quality-control techniques, e.g., majority voting introduced in Chap. 3 for aggregating answers from multiple crowd workers.

Equipped with this query language design, Qurk is flexible for requesters to customize their own task types and, on the other hand, also has higher requirements for the requesters, e.g., understanding HTML codes and the Qurk syntax.

6.2.3   Deco

The Deco system [14] puts more emphasis on crowdsourced data collection. In order to facilitate its query optimization approach (which will be introduced in Sect. 6.3.3), it designs mechanisms on the data model and query semantics.

- *Data model*: Deco separates requester and system views to ease crowdsourced data collection, as depicted in Fig. 6.5a. On one hand, it allows requesters or schema designers to define *concept relations*, which logically organize schema of the data to be collected. More specifically, it differentiates attributes in a relation into (1) *anchor attributes* which play roles similar to *primary keys* in traditional RDBMS and (2) *dependent attribute groups* which can be considered as the "properties" of tuples. Note that the design of anchor attributes and dependent attribute groups can specify the dependency in the process of data collection (see more details in Sect. 6.3.3). On the other hand, *raw schema* is for the data tables actually stored in the underlying RDBMS and is invisible to the schema designer and users. Normally, Deco stores anchor and dependent attributes separately in different relational tables.
- *Fetch semantics*: Deco designs *fetch rules* that allow schema designers to specify how data is collected from the crowd. The semantic of a fetch rule is get a value of attribute A_2 given the value of attribute A_1. This is formally presented as the form $A_1 \rightarrow A_2 : P$, where A_1 is a set of attributes used as *fetch conditions*, A_2 is the attribute to be fetched, and P is the fetch procedure for generating crowdsourcing task UIs. Given a fetch rule, Deco presents the values of attributes in A_1 and asks the crowd to give the values of attributes in A_2. Figure 6.5b provides three examples of fetch rules, asking for a new `car` tuple, filling values of `door-num` given existing car tuples, and filling `style` values. In particular, if A_1 is empty, the system fetches new values of attributes in A_1 and A_2.
- *Resolution rules*: As inconsistency may exist in the collected data, Deco can also specify resolution rules such as deduplication and majority voting to resolve inconsistencies in the collected data. This is similar to the **Combiner** keyword in Qurk. Deco incorporates the so-called resolution rules for aggregating inputs from the crowd. The current version of Deco focuses on two kinds of resolution rules, as shown in Fig. 6.5b: (1) *majority-of-3* that utilizes majority voting from a group of three crowd workers, and (2) *dumElim* that eliminates duplicated tuples from crowd collections. Obviously, the former rule is usually used for filling semantics, while the latter works for collecting tuples.

Based on the above fetch and resolution rules, Deco allows requesters to write SQL statements with a newly introduced keyword **MINTUPLES**. The usage of **MINTUPLES** is similar to that of the LIMIT keyword in SQL. However, the semantics are different: **MINTUPLES** n asks for at least n tuples from the crowd, while LIMIT n retrieves at most n tuples from relational tables. Figure 6.5c provides an example query that collects at least 8 "SUV" cars.

6.2.4 CDAS

The CDAS system [4] is also built on the relational data model. Different from the other crowdsourcing database systems, CDAS utilizes standard SQL as its

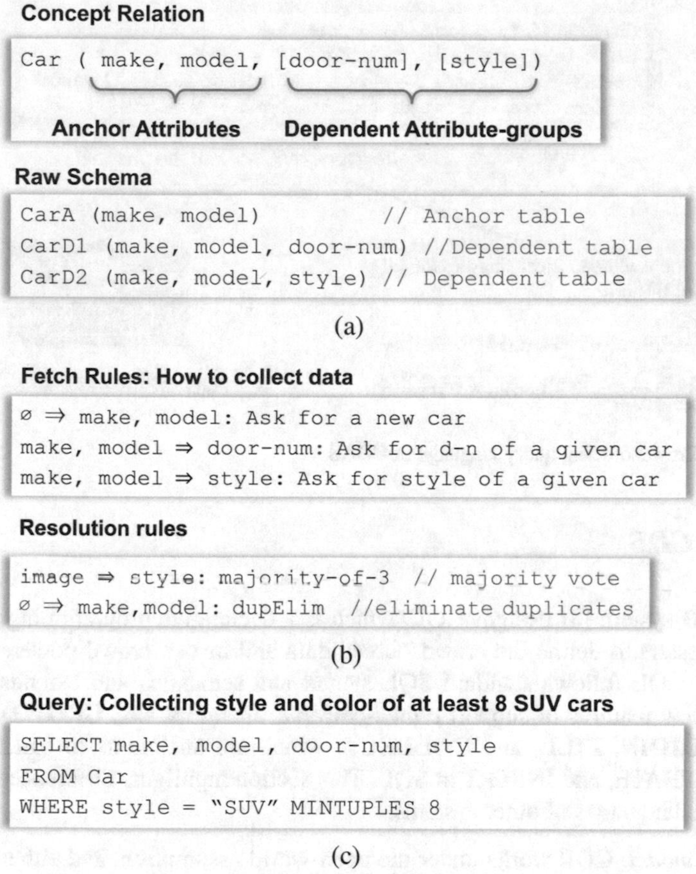

Concept Relation

```
Car ( make, model, [door-num], [style])
```

Anchor Attributes   Dependent Attribute-groups

Raw Schema

```
CarA (make, model)          // Anchor table
CarD1 (make, model, door-num) //Dependent table
CarD2 (make, model, style) // Dependent table
```

(a)

Fetch Rules: How to collect data

```
∅ ⇒ make, model: Ask for a new car
make, model ⇒ door-num: Ask for d-n of a given car
make, model ⇒ style: Ask for style of a given car
```

Resolution rules

```
image ⇒ style: majority-of-3  // majority vote
∅ ⇒ make,model: dupElim  //eliminate duplicates
```

(b)

Query: Collecting style and color of at least 8 SUV cars

```
SELECT make, model, door-num, style
FROM Car
WHERE style = "SUV" MINTUPLES 8
```

(c)

Fig. 6.5 Crowdsourcing query language of deco. (**a**) Data model: Concept relation vs. raw schema. (**b**) Fetch semantics for data collection. (**c**) An example query in deco

query language, instead of introducing a new query language. In order to support crowdsourcing on demand, CDAS allows tuple attributes to be marked as *unknown* by the schema designer a priori, and these attributes can be crowdsourced when evaluating SQL queries. Figure 6.6 illustrates an example query for CDAS, where values of both car_review.sentiment and car_image.color are unknown before executing crowdsourcing and can be easily recognized by the crowd. Thus, when evaluating the query, CDAS detects the unknown attributes and decides a better way of crowdsourcing. Note that, by default, CDAS evaluates predicates in the WHERE clause using crowdsourcing if the predicates involve attributes with unknown values. It also allows users to specify the predicates that they want to evaluate using only the values stored in the database by a **ByPass** keyword.

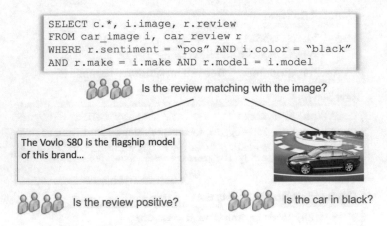

```
SELECT c.*, i.image, r.review
FROM car_image i, car_review r
WHERE r.sentiment = "pos" AND i.color = "black"
AND r.make = i.make AND r.model = i.model
```

Is the review matching with the image?

The Vovlo S80 is the flagship model of this brand...

Is the review positive?   Is the car in black?

Fig. 6.6 Crowdsourcing query language of CDAS

6.2.5 CDB

The CDB system [8] designs CQL, which is a declarative programming language for requesters to define the crowdsourced data and invoke crowd-powered manipulations. CQL follows standard SQL syntax and semantics and extends SQL by adding new features to support *crowd-powered* operators, i.e., **CROWDEQUAL**, **CROWDJOIN**, **FILL**, and **COLLECT**, which are analogous to SELECTION, JOIN, UPDATE, and INSERT in SQL. This section highlights differences between CQL and languages of other systems.

- *Data model:* CQL works under the open-world assumption, and either columns or tables can be crowdsourced by introducing a keyword **CROWD**, which is similar to the CrowdDB system [6]. Nevertheless, unlike CrowdDB, CQL does not have the requirement that *a **CROWD** table must have a primary key to determine if two workers contribute the same tuple*, because this is often not practical. Instead, CDB applies crowdsourced entity resolution techniques (see Sect. 7) for the collected tuples or records.
- *Crowd-powered collection*: CDB introduces two built-in keywords for data collection: **FILL** and **COLLECT**. The former operator asks the crowd to fill the values of a column, while the latter asks the crowd to collect a table. Figure 6.7 shows two examples of filling values of make for tuples in car_image and collecting "SUV" cars. One practical issue in **FILL** and **COLLECT** is the cleansing of the crowd-collected data, in particular, the entity resolution problem. In CDB, this issue is solved by an autocompletion interface and native crowdsourcing methods. First, CDB provides an *autocompletion* interface to the crowd workers to choose an existing value of an attribute, e.g., first typing a character "v" and then choosing "Volvo" from a suggested list. If no existing value fits a worker's tuple, the worker can also input a new value, which will

be added to the existing value set. This type of interface can reduce the variety of collected tuples or attributes, since workers could choose from the same set of values or learn how to represent new values. Second, there may still exist the cleansing issue, even if the aforementioned autocompletion interface is used. CDB then solves this problem in a crowdsourcing manner, such as using crowdsourced entity resolution techniques, by leveraging the query semantics of CQL.

- CQL defines crowdsourced operations, **CROWDEQUAL** and **CROWDJOIN** to solicit the crowd to perform filtering and join on crowdsourced or ordinary attributes. An example query is shown in Fig. 6.7 that joins tuples from car_image and car_review as well as filtering images containing "red" cars.

Moreover, CQL has a feature of introducing a *budget* mechanism to allow requesters to configure the cost constraint of crowdsourcing. On the one hand, with respect to the collection semantics, it is often unclear how many tuples or values can be collected due to the open-world assumption. Thus, a budget should be naturally introduced to bound the number of **COLLECT** or **FILL** tasks. On the other hand, with respect to the query semantics, as the amount of data may be huge, the requester often wants to set a budget to avoid running out of money when evaluating a CQL query. To achieve this goal, CQL introduces a keyword BUDGET, which can be attached to either collection or query semantics to set the number of tasks. The requester only needs to provide the budget, and CDB query optimization, together with plan generation components, will design algorithms to fully utilize the budget for producing better results.

Fill Semantics

```
FILL car_image.color
WHERE car_image.make = "Volvo";
```

Collect Semantics

```
COLLECT car.make, car.model
WHERE car.style = "SUV";
```

Query Semantics

```
SELECT *
FROM car_image M, car C, car_review R
WHERE M.(make,model) CROWDJOIN C.(make,model)
AND R.(make, model) CROWDJOIN C.(make,model)
AND M.color CROWDEQUAL "red"
```

Fig. 6.7 Crowdsourcing query language of CDB

6.3   Crowdsourcing Query Optimization

This section introduces query processing and optimization techniques in existing crowdsourcing database systems. Typically, the basic idea of query processing in crowdsourcing database systems is to develop crowd-powered operators (or crowd operators for simplicity). Similar to traditional RDBMS, these crowd operators will form a query plan to evaluate a query, and optimization techniques are introduced to choose the best one from the possible query plans. However, the key difference is that each of these crowd operators abstracts a specific type of operation that can be processed by the crowd. More specifically, these operators are initialized with predefined UI templates and some standard HIT parameters used for crowdsourcing. At runtime, a crowd operator consumes a set of tuples and instantiates crowd-sourcing tasks by filling values of the tuples into the corresponding UI templates. To implement these crowd operators, existing systems have introduced their own schemes, which will be presented in this chapter.

Moreover, the performance of query processing in crowdsourcing database systems is measured by multiple metrics, i.e., quality, cost, and latency. Thus, existing crowdsourcing database systems invent mechanisms to control these metrics so as to achieve superior query optimization.

6.3.1   CrowdDB

In query processing, CrowdDB [6] introduces three crowd operators.

- **CrowdProbe**: Collect missing information of attributes or new tuples from the crowd. The typical user interface of **CrowdProbe** is a form with several fields for collecting information from the crowd.
- **CrowdJoin**: Implement an index nested-loop join over two tables, where at least one table is crowdsourced. In particular, the inner relation must be a **CROWD** table (see Sect. 6.2.1), and the user interface is used to crowdsource new tuples of the inner relation which can be joined with the tuples in the outer relation.
- **CrowdCompare**: This operator is designed to implement two keywords, **CROWDEQUAL** and **CROWDORDER**, defined in CrowdDB's query language CrowdSQL. The interface of the operator crowdsources two tuples and leverages the crowd to compare these tuples. **CROWDEQUAL** compares two values and asks the crowd to decide whether they have the same value. **CROWDORDER** asks the crowd to give an order according to a predefined attribute.

Figure 6.8 shows examples of these crowd operators. Operators **CrowdProbe** and **CrowdJoin** are illustrated in Fig. 6.8a: **CrowdProbe** is used to collect new car tuples with make "Volvo" from the crowd, while **CrowdJoin** implements a join over car and car_review, collecting missing reviews on demand. Operator

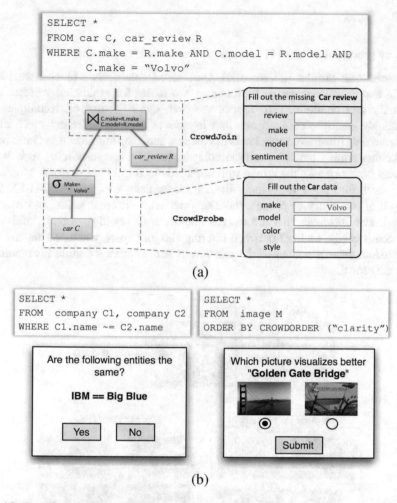

Fig. 6.8 CrowdDB, crowdsourcing query operators (**a**) Operators for crowdsourcing missing data (**b**) Operators for crowdsourced comparison

CrowdCompare is used to support evaluating **CROWDEQUAL** (e.g., is "IBM == Big Blue") and **CROWDORDER** (e.g., which picture visualizes better "Golden Gate Bridge").

For query optimization, CrowdDB devises rule-based optimization techniques for processing queries with multiple operators, such as pushing down selection predicates and determining join order.

6.3.2 Qurk

In query processing, Qurk [11] focuses on implementing join and sort.

- **CrowdJoin**: Similar to CrowdDB, Qurk also implements a block nested-loop join and crowdsources the tuples from two tables for evaluating whether they satisfy join conditions. To reduce the cost, Qurk focuses on techniques for batching multiple comparisons and introduces a feature-filtering mechanism. Figure 6.9a illustrates three kinds of batching schemes introduced in Qurk: naive batching simply batches multiple comparisons in a crowdsourcing task, while smart batching asks the crowd to link records to fulfill comparisons. Moreover, Fig. 6.9b illustrates the feature-filtering mechanism, e.g., the **POSSIBLY** keyword allows Qurk to prune candidate pairs with different `make` and `style`. Qurk also discusses that the feature filtering may not always be helpful, as it depends on the tradeoff between filtering and join costs, whether a feature will introduce false negatives to the join results and whether a feature is ambiguous to the crowd.

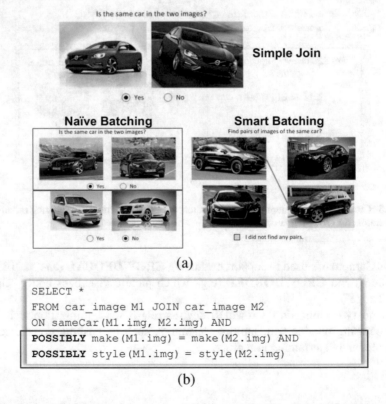

(a)

```
SELECT *
FROM car_image M1 JOIN car_image M2
ON sameCar(M1.img, M2.img) AND
POSSIBLY make(M1.img) = make(M2.img) AND
POSSIBLY style(M1.img) = style(M2.img)
```

(b)

Fig. 6.9 Qurk: Crowdsourcing join query optimization (**a**) Optimization with batching multiple comparisons (**b**) Optimization with feature filtering

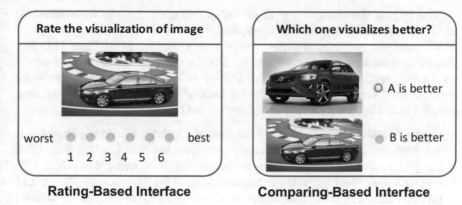

Fig. 6.10 Qurk: Optimization for crowdsourced sort design

- **CrowdSort**: Qurk implements two basic approaches to execute sort, as shown in
  Fig. 6.10. The comparison-based approach solicits the crowd to directly specify
  the ordering of items. This approach may be expensive for large datasets due to
  the quadratic comparison. Another task type, rating, is used to reduce the cost
  using a well-defined interface (see Sect. 7.4).

  Qurk also has two important components for cost optimization: task cache and
  task model. Task cache maintains the crowdsourced answers from previous tasks,
  and task model trains a model to predict the results for the tasks based on the
  data that are already collected from the crowd. So if one task can get necessary
  information from the task cache or task model, it will not be published to the
  crowdsourcing platform. Once the task cannot get useful information from the
  task cache or task model, it will be pushed to the task compiler. The compiler
  generates and publishes the tasks to the crowdsourcing platform. A statistic manager
  determines the number of tasks, assignments, and the cost for each task.

6.3.3 Deco

Deco [14] focuses on crowdsourcing missing values or new tuples based on the
defined *fetch* rules, as introduced previously in Sect. 6.2.3. To this end, Deco designs
the **Fetch** operator, which can be instantiated as crowdsourcing tasks with text fields
for inputing missing values or tuples. Moreover, Deco also supports other machine-
based operators, such as dependent left outer join, filter, and scan, for processing the
fetched data.

Based on the defined operators, given a complicated query, a fundamental query
optimization problem defined in [15] is as follows:

How to find the best query plan for the query, which has the least estimated monetary cost across all possible query plans?

To solve the problem, the query optimization component of Deco [15] first defines the monetary cost. Considering the fact that the existing data in the database can be leveraged, the cost is formally defined as the new data that needs to be obtained from the crowd. In order to find the best query plan with the minimum cost, the following two problems are addressed in [15].

- **Cost estimation**: How to estimate the cost of a query plan. As a query plan is executed, the database may collect new data from the crowd, which may affect the cost estimation of subsequent processes. By considering this effect, [15] proposes an iterative approach to estimate the cost for a query plan. Figure 6.11 provides an example of cost estimation. Consider an example database with an anchor table `CarA` and a dependent table `CarD2`, as well as some statistics shown in Fig. 6.11a. The cost of the query plan in Fig. 6.11b can be estimated in the following way:

(a) Consider operator **Resolve** with a target of fetching 8 "SUV" cars. This operator examines the current status of the database and finds 2 "SUV", 1 "Sedan," and 1 car with unknown style information. As the selectivity of "SUV" cars is 0.1, it estimates that the database has an expected number 2.1 of "SUV" cars. Based on this, it specifies a new target $8 - 2.1 = 5.9$ "SUV" cars to its descendant **fetch** operator.

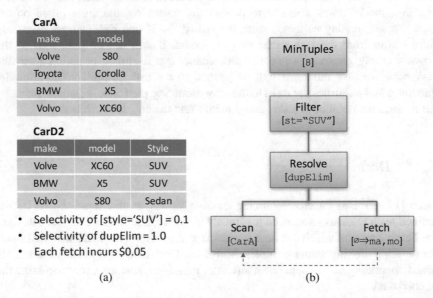

(a)   (b)

Fig. 6.11 Deco: Cost-based crowdsourcing query optimization. (**a**) An example database (**b**) Cost estimation

(b) The **fetch** operator, given a target of 5.9 "SUV" cars, estimates that it needs to fetch at least 59 cars from the crowd, because the selectivity of "SUV" cars is 0.1. Based on this, it instantiates crowdsourcing tasks to fetch data.

- **Optimal query plan generation**: Simply enumerating all possible query plans is computationally expensive and [15] considers to reuse the common subplans in order to reduce the redundant computation. Then the best query plan, i.e., with least estimated cost, can be returned. Figure 6.12 compares two query plans. The first one shown in Fig. 6.12a first fetches car tuples and then filters the ones with style "SUV." On the contrary, the second query plan, which is also called *reverse plan* as shown in Fig. 6.12a, utilizes "SUV" as a condition of fetching new tuples and performs better than the first query plan.

6.3.4 CDAS

CDAS [4] focuses on optimizing selection-join (SJ) queries. To this end, it has introduced the following three crowd operators.

- To support selection query semantics, CDAS implements operator **CrowdSelect** (CSelect) to solicit the crowd to filter tuples satisfying certain conditions. A typical user interface of CrowdSelect presents a tuple and some selection conditions and asks the crowd to judge whether the tuple satisfies the conditions.
- To support the join query semantics, CDAS provides a two-pronged solution that implements two operators, **CrowdJoin** (CJoin) and **CrowdFill** (CFill). CJoin presents two tuples, each from a relation, and asks the crowd to directly compare the tuples to be joined. On the other hand, CFill crowdsources the missing values in tuples' attributes to be joined and then uses traditional RDBMS techniques to join the tuples. In particular, a user interface of CFill provides a list of candidate values from the attribute value's domain for the crowd to choose or asks the crowd to contribute a new value if no value in the list can be filled to the tuple.

A highlight in CDAS query optimization is to consider the cost-latency trade-off when evaluating a crowdsourcing query. Recall that *cost* and *latency*, respectively, measure the money and time spent for the crowdsourcing process. In CDAS, cost is measured by the total money spent for evaluating all operators in a query plan, while latency is measured by the "round" of crowdsourcing, which is naturally equivalent to the *height* of a query plan tree. Figure 6.13 provides two example query plans: the sequential plan in Fig. 6.13a takes *four* rounds, each of which evaluates a selection condition (e.g., style ="Sedan"), while the parallel plan in Fig. 6.13b only takes *one* round. Obviously, in terms of latency, the second plan is better. On the contrary, the sequential plan would incur lesser cost than the latter, as tuples would be filtered in the bottom operators, which leads to less input to the top operators. Based on this observation, CDAS introduces two objectives for query optimization.

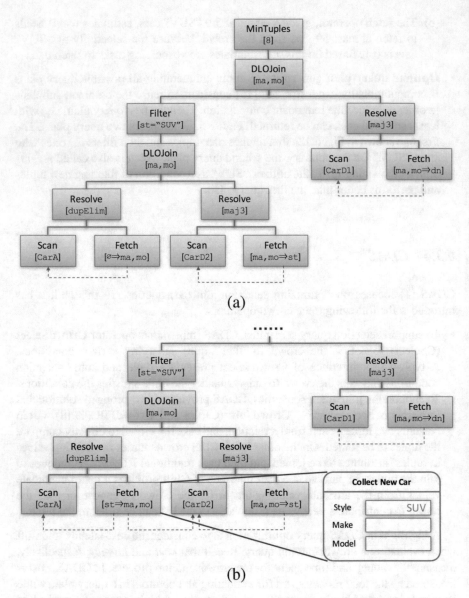

Fig. 6.12 Deco: Alternative query optimization plans (**a**) Initial crowdsourcing query plan (**b**) Reverse crowdsourcing query plan

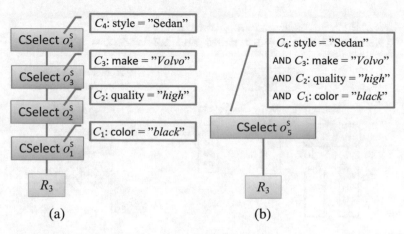

Fig. 6.13 CDAS, cost-latency trade-off in query optimization (**a**) Sequential query plan (**b**) Parallel query plan

- **Cost minimization:** Find a query plan that minimizes the monetary cost
- **Cost bounded latency minimization:** Find a query plan with bounded cost and the minimum latency

CDAS focuses on developing techniques for optimizing queries based on the aforementioned two objectives.

- Selection query optimization as shown in Fig. 6.13 is relatively trivial: cost minimization can be achieved by sorting operators by selectivity, and cost bounded latency minimization can be solved by a dynamic programming algorithm.
- Join query optimization is more interesting. CDAS introduces a hybrid CFill-CJoin framework that consist of two steps, as illustrated in Fig. 6.14. The first step utilizes CFill operators to crowdsource missing values on some join attributes, which is illustrated as the intermediate nodes of the left tree in Fig. 6.14: it first fills the make for all tuples and then fills the style of the tuples with make being "Volvo." The second step is to use CJoin to directly compare the tuples with the same filled attributes, as illustrated in the leaf nodes of the left tree in Fig. 6.14. The framework can be represented as the query plan in the right part of Fig. 6.14, which consists of the CFill operators on the bottom and CJoin on the top. As both CFill and CJoin incur cost and latency, the key in join query optimization is to determine a best CFill plan, i.e., the left tree in the figure. CDAS proves that finding such a plan with minimum overall cost is NP-hard and devises heuristic algorithms to solve the problem.

Based on the techniques for optimizing selections and joins, CDAS further considers the optimization of a general selection-join (SJ) query. For the objective of cost minimization, CDAS can optimize an SJ query similarly to traditional RDBMS: pushing down selections, determining join ordering, and then invoking the aforementioned techniques for each selection/join operator. However, optimization

```
SELECT * FROM car R2, car_image R3
WHERE R2.make = R3.make AND R2.model = R3.model
```

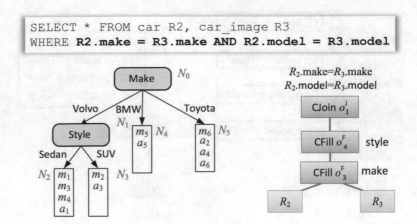

Fig. 6.14 CDAS: A fill-join hybrid join optimization

Fig. 6.15 CDAS: Latency constraint allocation for query optimization

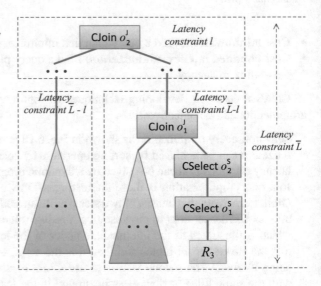

with the objective of cost bounded latency minimization is nontrivial, as one has to carefully allocate the "latency constraints" across operators in the entire plan, as illustrated in Fig. 6.15: allowing CSelect to use *two* rounds and CJoin to use *one* round may be very different from the allocation of *one* round for CSelect and *two* rounds for CJoin. CDAS devises a dynamic programming algorithm to solve this problem.

Fig. 6.16 CDB: Graph-based query processing model

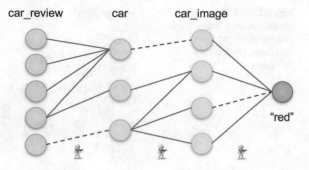

6.3.5   CDB

So far, all of the crowdsourcing database systems adopt the traditional relational query model that formalizes a query as a treelike query plan, which aims at selecting an optimized table-level join order to optimize a query. Different from them, CDB [8] defines a graph-based query model, in order to provide a fine-grained optimization on the crowdsourced data. Given a CQL query, CDB constructs a graph, where each vertex is a tuple of a table in the CQL and each edge connects two tuples based on the join/selection predicates in the CQL, and this graph can be used to provide tuple-level optimization. More specifically, each edge between two tuples is crowdsourced as a task that asks the crowd whether the tuples can be joined. Then, the crowdsourcing results are annotated in the graph, as illustrated in Fig. 6.16: a blue solid (red dotted) edge denotes that the two tuples can (cannot) be successfully joined. Based on this, given a CQL query with N join predicates, an answer of the query is a path consisting of $N + 1$ blue solid edges. CDB can support crowdsourced selection, join, aggregation, sort, and top-k.

Based on this graph-based formalization, CDB considers query optimizing in all three dimensions, cost, latency, and quality.

- **Cost control** The graph-based model offers more potential for minimizing crowdsourcing cost. Observed from Fig. 6.17a, the tree model determines the best join ordering that first asks *five* tasks between car and car_image, then *two* tasks between car_review and car, and finally *one* task between car_image and a predicate color ="red." Overall, this invokes *eight* tasks. On the other hand, a graph-based model only needs to find which red-dotted edges are more beneficial to prune other edges. For example, as shown in Fig. 6.17a, one can first ask the first edge between car and car_image and annotate it as red dotted, so as to prune all the edges related to this edge. Then, it crowdsources other edges having effective pruning powers. In this way, the graph-based model only needs to ask *five* tasks. CDB proves that the optimization problem of determining which edges to ask is NP-hard and discusses various heuristic algorithms, such as greedy, expectation-based, etc., to solve the problem.

Fig. 6.17 CDB: Crowdsourcing query optimization. (**a**) Cost reduction. (**b**) Latency reduction

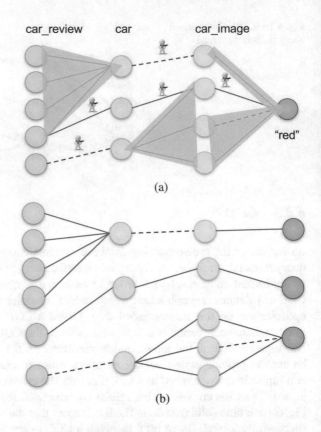

- **Latency control** Similar to CDAS, CDB also applies the *round*-based model to measure the latency. The basic idea of latency control in CDB is as follows. Given two edges e and e', CDB checks whether they are in the same candidate (i.e., a path that could be an answer). If they are in the same candidate, it labels them as "conflict," because asking an edge may prune other edges; otherwise they are non-conflict. CDB asks non-conflict edges simultaneously but cannot ask conflict edges. For example, consider two conflict edges. If an edge is colored red, then the other edge does not need to be asked. To check whether two edges are in the same candidate, one can enumerate all the candidates of an edge, e.g., e, and check whether they contain e. If yes, they are in the same candidate and no, otherwise. However this method is rather expensive. Thus, CDB proposes several effective rules based on connected components to detect whether two edges can be asked simultaneously.
- **Quality control** In order to derive high-quality results based on workers' answers, it is important to do quality control. CDB controls quality at two time stamps: (1) when a worker answers a task, it estimates the worker's quality and infers the truth of the answered task, called "truth inference"; (2) when a worker comes and requests for new tasks, it considers the worker's quality and

assigns tasks with the highest improvement in quality to the worker, called "task assignment." More technical details of truth inference and task assignment can be found in Chap. 3. Moreover, CDB supports four types of tasks: single-choice, multiple-choice, fill in the blank, and collection tasks.

CDB can simultaneously optimize the cost, latency, and quality using the graph model.

6.4 Summary of Crowdsourcing Database Systems

Figure 6.18 summarizes the existing crowdsourcing database systems.

- **Optimization models** Query optimization in the existing systems can be classified into rule based and cost based. CrowdDB [6] uses rule-based optimization, e.g., pushing down selection predicates and determining join order, which may not be able to find the best query plan with the lowest cost. Other systems [4, 11, 13] design cost models that aim to find the query plan with the minimum cost. However, these systems still adopt a tree model that selects an optimized table-level join order to optimize the query. As analyzed above, the tree model gives the same order for different joined tuples and limits the optimization potential that different joined tuples can be optimized for different orders. Different from the above, CDB devises graph-based query optimization to perform fine-grained tuple-level optimization, which has the potential to save a huge amount of cost.
- **Optimization objectives** Crowdsourcing query optimization should consider trade-offs among cost, latency, and quality, because any single-objective optimization, such as smaller cost with lower quality, higher quality with larger latency, etc., is not desirable. As shown in Fig. 6.18, most existing systems optimize monetary cost, utilize majority voting (MV) for quality control, and do not optimize latency. CDAS [4] optimizes latency by simply considering data dependencies. CDB develops techniques based on data inference to reduce latency in a more effective way. Considering the quality concern, most of the systems leverage existing majority voting or its variants, which is only applicable in single-choice tasks, while CDB also takes quality into consideration and devises more sophisticated quality-control strategies (i.e., truth inference and task assignment) for either single-choice, multiple-choice, fill in the blank, and collection tasks.
- **Optimized crowd operators** Let us consider commonly used crowd-powered operators and examine whether they are optimized in the existing systems, as shown in Fig. 6.18. CrowdDB [6] optimizes **SELECT**, **JOIN**, **COLLECT**, and **FILL**. Qurk [11] focuses on crowd-powered **SELECT** and **JOIN**. Deco [13] considers more on **FILL** and **COLLECT** (i.e., the *fetch* operator in Deco) while also supporting **SELECT** and **JOIN**. CDAS [4] only supports **SELECT**, **JOIN**, and **FILL** operators. CDB optimizes all of the operators by introducing a crowd-sourcing query language, which can fulfill more crowdsourcing requirements.

		CrowdDB	Qurk	Deco	CDAS	CDB
Optimized Crowd Operators	COLLECT	√	×	√	×	√
	FILL	√	×	√	√	√
	SELECT	√	√	√	√	√
	JOIN	√	√	√	√	√
Optimization Objectives	Cost	√	√	√	√	√
	Latency	×	×	×	√	√
	Quality	MV	MV	MV	MV	√
Optimization Strategies	Cost-Model	×	√	√	√	√
	Tuple-Level	×	×	×	×	√
	Budget-Supported	×	×	×	×	√
Task Deployment	Cross-Market	×	×	×	×	√

Fig. 6.18 Comparison of crowdsourcing database systems

- **Task deployment** Most of the systems support publishing Human Intelligence Tasks (HITs) only on one individual crowdsourcing market, such as AMT [1], and the results may be affected by the bias of the market. CDB has the flexibility of cross-market HIT deployment by simultaneously publishing HITs to AMT [1], CrowdFlower [2], etc.

References

1. Amazon mechanical turk. https://www.mturk.com/
2. Crowdflower. http://www.crowdflower.com
3. Davidson, S.B., Khanna, S., Milo, T., Roy, S.: Using the crowd for top-k and group-by queries. In: ICDT, pp. 225–236 (2013)
4. Fan, J., Zhang, M., Kok, S., Lu, M., Ooi, B.C.: Crowdop: Query optimization for declarative crowdsourcing systems. IEEE Trans. Knowl. Data Eng. **27**(8), 2078–2092 (2015)
5. Feng, A., Franklin, M.J., Kossmann, D., Kraska, T., Madden, S., Ramesh, S., Wang, A., Xin, R.: Crowddb: Query processing with the vldb crowd. PVLDB **4**(12), 1387–1390 (2011)
6. Franklin, M.J., Kossmann, D., Kraska, T., Ramesh, S., Xin, R.: Crowddb: answering queries with crowdsourcing. In: SIGMOD, pp. 61–72 (2011)
7. Guo, S., Parameswaran, A.G., Garcia-Molina, H.: So who won?: dynamic max discovery with the crowd. In: SIGMOD, pp. 385–396 (2012)
8. Li, G., Chai, C., Fan, J., Weng, X., Li, J., Zheng, Y., Li, Y., Yu, X., Zhang, X., Yuan, H.: CDB: optimizing queries with crowd-based selections and joins. In: SIGMOD, pp. 1463–1478 (2017)
9. Marcus, A., Karger, D.R., Madden, S., Miller, R., Oh, S.: Counting with the crowd. PVLDB **6**(2), 109–120 (2012)

10. Marcus, A., Wu, E., Karger, D.R., Madden, S., Miller, R.C.: Human-powered sorts and joins. PVLDB **5**(1), 13–24 (2011)
11. Marcus, A., Wu, E., Madden, S., Miller, R.C.: Crowdsourced databases: Query processing with people. In: CIDR, pp. 211–214 (2011)
12. Parameswaran, A.G., Garcia-Molina, H., Park, H., Polyzotis, N., Ramesh, A., Widom, J.: Crowdscreen: algorithms for filtering data with humans. In: SIGMOD, pp. 361–372 (2012)
13. Parameswaran, A.G., Park, H., Garcia-Molina, H., Polyzotis, N., Widom, J.: Deco: declarative crowdsourcing. In: CIKM, pp. 1203–1212. ACM (2012)
14. Park, H., Pang, R., Parameswaran, A.G., Garcia-Molina, H., Polyzotis, N., Widom, J.: Deco: A system for declarative crowdsourcing. PVLDB **5**(12), 1990–1993 (2012)
15. Park, H., Widom, J.: Query optimization over crowdsourced data. PVLDB **6**(10), 781–792 (2013)
16. Park, H., Widom, J.: Crowdfill: collecting structured data from the crowd. In: SIGMOD, pp. 577–588 (2014)
17. Sarma, A.D., Parameswaran, A.G., Garcia-Molina, H., Halevy, A.Y.: Crowd-powered find algorithms. In: ICDE, pp. 964–975 (2014)
18. Trushkowsky, B., Kraska, T., Franklin, M.J., Sarkar, P.: Crowdsourced enumeration queries. In: ICDE, pp. 673–684 (2013)
19. Wang, J., Kraska, T., Franklin, M.J., Feng, J.: CrowdER: crowdsourcing entity resolution. PVLDB **5**(11), 1483–1494 (2012)
20. Wang, J., Li, G., Kraska, T., Franklin, M.J., Feng, J.: Leveraging transitive relations for crowdsourced joins. In: SIGMOD, pp. 229–240 (2013)
21. Yan, T., Kumar, V., Ganesan, D.: Crowdsearch: exploiting crowds for accurate real-time image search on mobile phones. In: MobiSys, pp. 77–90 (2010)

Chapter 7
Crowdsourced Operators

To obtain high-quality results, different applications require the use of different crowdsourced operators, which have operator-specific optimization goals over three factors: cost, quality, and latency. This chapter reviews how crowdsourced operators (i.e., *crowdsourced selection, crowdsourced collection, crowdsourced join, crowdsourced sort, crowdsourced top-k, crowdsourced max/min, crowdsourced aggregation, crowdsourced categorization, crowdsourced skyline, crowdsourced planning, crowdsourced schema matching, crowd mining, spatial crowdsourcing*) can be implemented and optimized. Specifically, for each operator, we summarize its task type, the optimization goal (cost, quality, or latency), and the techniques used to achieve the goal.

7.1 Crowdsourced Selection

Given a set of items, crowdsourced selection selects the items that satisfy some specific constraints, e.g., selecting the images that contain both mountains and humans. Existing works on crowdsourced selection include three operators: crowdsourced filtering [65, 66], crowdsourced find [76], and crowdsourced search [100]. Their difference is the *targeted size* of returned results:

Definition 7.1 (Crowdsourced Filtering) Crowdsourced filtering [65, 66] (or all selection) returns *all* items that satisfy the given constraints.

Definition 7.2 (Crowdsourced Find) Crowdsourced find [76] (or *k*-selection) returns *a predetermined number of (denoted as k)* items that satisfy the given constraints.

Definition 7.3 (Crowdsourced Search) Crowdsourced search [100] (or 1-Selection) returns *only one* item that satisfies the given constraints.

© Springer Nature Singapore Pte Ltd. 2018
G. Li et al., *Crowdsourced Data Management*,
https://doi.org/10.1007/978-981-10-7847-7_7

For example, given 100 images, suppose there are 20 images that have both mountains and humans. Crowdsourced filtering selects all 20 targeted images, crowdsourced find requires to select a bounded number of (say, five) targeted images, and crowdsourced search only selects one targeted image.

The remainder of this section, respectively, introduces how existing works deal with these three crowdsourced operators.

7.1.1 Crowdsourced Filtering

Parameswaran et al. [66] first focus on a simplified crowdsourced filtering problem with only a single constraint and then extend their solution to multiple constraints. A typical user interface of the task type used by their work is to ask the crowd whether or not an item satisfies a given constraint (e.g., "Does the following image contain SUV cars?"), and the possible answer is Yes or No.

Parameswaran et al. [66] assign such a task to multiple workers and propose a *strategy* function to balance the trade-off between quality and cost. The strategy function takes the answers collected for each task (the number of Yes/No) as input and outputs one of the following decisions: (1) the item satisfies the given constraint, (2) the item does not satisfy the given constraint, and (3) the item should be asked again. For an item, the cost is defined as the number of workers answering its corresponding task and the quality is defined as the probability of correctly answering the item. Figure 7.1 provides an example of filtering images containing SUV cars. Each two-dimensional point (x, y) depicted in the right part of the figure represents a status of collected answers. For example, the green point in the figure represents that 1 Yes answer and 2 No answers have been collected. Given such a status, a strategy needs to make a decision from the following three choices: (1) outputting the filtering result "No," (2) outputting the filtering result "Yes," and (3) continuing to ask a new task for reaching point $(2, 2)$ or $(1, 3)$. From the example, one can see that the cost of each point can be naturally defined as the number of tasks, i.e., $x + y$. In addition, if a point has outputted a Yes/No answer, it serves as a "terminated" status, and Parameswaran et al. estimate the probability of the output being correct for each of such points.

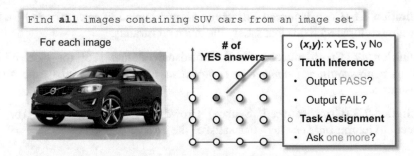

Fig. 7.1 An example of crowdsourced filtering

Based on the strategy's definition, Parameswaran et al. study how to find an optimal strategy that satisfies cost and error constraints.

Definition 7.4 (The Problem with Cost and Error Constraints) Given an error threshold τ, find an optimal strategy that minimizes the expected cost w.r.t. the error constraint, i.e., the expected error $< \tau$.

To solve the problem, Parameswaran et al. propose a brute-force algorithm for the problem and then develop an efficient approximate algorithm that performs well in practice. They also extend the output of a strategy to a probability distribution (rather than a single choice) and study how to find an optimal probabilistic strategy. In their work [66], they make two assumptions: (1) all workers are of equal quality to answer a given task, and (2) all items are of equal difficulty to satisfy a given constraint. They also relax these assumptions in a follow-up paper [65], which addresses crowdsourced filtering query by considering that (1) workers may have different qualities in answering a given task and (2) certain items are easier (or harder) to satisfy a given constraint. By using the same task type as [66], i.e., asking the crowd whether or not an item satisfies a given constraint, [65] seeks to find an optimal strategy by considering the trade-off between cost and quality.

Weng et al. [97] studied how to optimize a query with multiple filtering constraints, e.g., selecting the female photos with dark eyes and wearing sunglasses. A straightforward method asks the crowd to answer every entity by checking every predicate in the query. Obviously, this method involves huge monetary cost. Instead, they select an optimized predicate order and ask the crowd to answer the entities following the order. Since if an entity does not satisfy a predicate, one can prune this entity without needing to ask other predicates, and thus this method can reduce the cost. There are two challenges in finding the optimized predicate order. The first is how to detect the predicate order, and the second is to capture correlation among different predicates. To address this problem, Weng et al. propose a predicate order-based framework to reduce monetary cost. Firstly, they define an expectation tree to store selectivities on predicates and estimate the best predicate order. In each iteration, they estimate the best predicate order from the expectation tree and then choose a predicate as a question to ask the crowd. After getting the result of the current predicate, they choose next predicate to ask until getting the result. The expectation tree will be updated using the answer obtained from the crowd and continue to the next iteration. They also study the problem of answering multiple queries simultaneously and reduce its cost using the correlation between queries.

7.1.2 Crowdsourced Find

As crowdsourced find cares more about how fast k qualified items can be selected, Sarma et al. [76] mainly focus on the trade-off between cost and latency. Specifically, they define latency as the number of rounds (see Sect. 5.4) and cost as the number of tasks asked. A typical user interface of the task type used is the same as that in crowdsourced filtering, i.e., to ask if an item satisfies a given constraint.

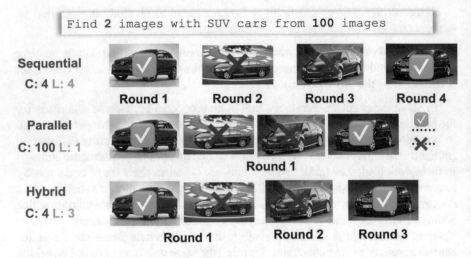

Fig. 7.2 An example of crowdsourced find

Assuming that workers always give true answers, Sarma et al. study a sequential algorithm that asks one task at each round and stops if k qualified items are observed. It is easy to see that the sequential algorithm requires the least cost. Using the sequential algorithm as a baseline, they further study how to improve its latency without increasing the cost.

To compare the sequential algorithm and its alternative strategies, consider the example shown in Fig. 7.2. Suppose to find 2 target images with SUV cars from 100 images. There may be the following strategies to achieve this goal.

- **Sequential strategy**: The strategy crowdsources the images one by one until *two* target images are found. In our example, both the cost (# of crowdsourced tasks) and the latency (# of rounds) of the strategy are 4.
- **Parallel strategy**: This strategy crowdsources all of the 100 images in parallel and achieves the goal of finding 2 target images in only 1 round. However, the cost of this strategy is 100 tasks.
- **Hybrid strategy**: As the goal is to find *two* target images, this strategy first crowdsources *two* images and collects the answer, i.e., *one* Yes and *one* No. Then, it modifies the goal to *one* and thus crowdsources *one* image, which is labeled by the crowd as No. Next, it crowdsources *one* image again and again until *one* "Yes" answer is obtained. In our example, this strategy crowdsources *four* tasks and takes *three* rounds.

Sarma et al. [76] develop an efficient algorithm that selects tasks for the next round based on the answers collected from the previous rounds. By relaxing the requirement that an algorithm has to reach the optimal cost, they study other problem formulations with additive and approximation error bounds on the cost constraint. Finally, they propose quality-control techniques to deal with the case that the workers may make errors.

7.1.3 Crowdsourced Search

Crowdsourced search, which requires only one qualified item to be returned, is specifically studied in [100]. It focuses on a real-world application: search a target image in real time on mobile phones. For example, given a target image with a building, and a set of candidate images, the query is to search an image in the candidate images that contains the same building. Each task is to put a candidate image and the target image together and ask workers whether the two images contain the same building. Workers will select Yes or No as the answer. Each search query has a deadline, and the objective of the query is defined as below.

Definition 7.5 (Crowdsourced Image Search) Given a target image and a deadline, select at least one correct image from a set of candidate images before the deadline while minimizing the cost.

The above objective considers the trade-off among three factors: cost, quality, and latency. The cost is defined as the number of answers collected. The quality is defined as the probability of correctly selecting one image out, where a majority-5 rule[1] is used to decide the result. The latency is predicted using a statistical model. Yan et al. [100] consider all these three factors and design an efficient and effective algorithm to meet the objective.

7.2 Crowdsourced Collection

Different from the abovementioned selection operators which select items from a given set of *known* items (closed-world assumption), crowdsourced collection tries to collect *unknown* items from the crowd (open-world assumption). Typical crowdsourced collection operators contain crowdsourced enumeration [84] and crowdsourced fill [68]. The former asks the crowd to enumerate a list of objects that satisfy a given constraint, and the latter asks the crowd to fill missing values in a table.

7.2.1 Crowdsourced Enumeration

Crowdsourced enumeration query [84] is defined as

Definition 7.6 (Crowdsourced Enumeration) A crowdsourced enumeration query utilizes the crowd to enumerate all items that satisfy a given constraint.

---

[1]Each task is assigned to five workers at most, and majority voting is used to aggregate answers, i.e., the result is returned upon getting three consistent answers.

An example of crowdsourced enumeration is to find all the state names in the USA, where each task with the description "Please give me one or more states in the US" is assigned to workers. If two workers' answers are (Texas, Utah) and (Utah, Arizona), respectively, then the returned result will be (Texas, Utah, Arizona). As workers' answers are collected, an important problem is: *When is the result set complete?*, that is, to estimate when the result size is equal to the total number of states, which is unknown in advance. To solve the problem, Trushkowsky et al. [84] consider a similar problem studied in biology and statistics, called *species estimation*, which counts the number of unique species of animals on an island by putting out traps each night, and in the next morning, the collected animals are counted and released, and then the process is repeated daily. However, directly using the estimator (which estimates the total number of species in the species estimation) may not well capture or even contradict human behavior. One typical difference is that species estimation problem samples animals *with replacement*, while in crowdsourcing, each worker "samples" their answers *without replacement* from some underlying distribution. Moreover, the underlying distributions of different workers may vary a lot, and the estimator in the species estimation cannot capture the worker skew and arrival rates. That is, if a worker (called streaker) arrives and suddenly dominates the number of answers, then the result size will be overestimated (Fig. 7.3).

Trushkowsky et al. [84] propose a new estimator to overcome the limitations. The estimator especially solves the overestimation issues caused by the existence of streakers. The way to find streakers is based on an observation that streakers give many unique answers (i.e., answers not provided by others). Trushkowsky et al. consider the effect of streakers in developing the new estimator. The developed estimator considers the trade-off between cost and quality, where cost is defined as the number of tasks asked and quality is defined as the completeness of the returned query results (w.r.t. the correct results). To get a better trade-off between cost and quality, Trushkowsky et al. develop a pay-as-you-go approach which predicts the benefit of getting additional unique answers by spending more budgets. If the benefit is not satisfactory by investing more budgets, then it can terminate asking more tasks, and the budgets can be saved. Chung et al. [19] extend the estimation

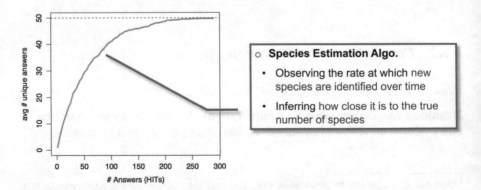

Fig. 7.3 Species estimation algorithm for crowdsourced enumeration

techniques to support aggregate queries, such as SUM, AVG, MAX/MIN, etc., by analyzing both coverage of the collected entities and the correlation between attributes and entities. Rekatsinas et al. [73] extend the underlying data model to a structured domain with hierarchical structure (e.g., restaurant with location and cuisine) and aim to maximize collection coverage under a monetary budget.

Fan et al. study *distribution-aware crowdsourced entity collection*: Given an expected distribution w.r.t. an attribute (e.g., region or year), it aims to collect a set of entities via crowdsourcing and minimize the difference of the entity distribution from the expected distribution [30]. Evidently, in many real scenarios, users may have *distribution requirements* on the entities collected from the crowd. For example, users often want the collected POIs (e.g., roadside parkings) to be evenly distributed in an area. However, each individual worker may have its own *bias* of data collection, leading to diverse distributions across different workers, as shown in Fig. 7.4a. To address the problem, they propose an adaptive worker selection approach to address this problem (Fig. 7.4b). The approach estimates underlying entity distribution of workers on-the-fly based on the collected entities. Then, it adaptively selects the best set of workers that minimizes the difference from the expected distribution. Once workers submit their answers, it adjusts the estimation of workers' underlying distributions for subsequent adaptive worker selections.

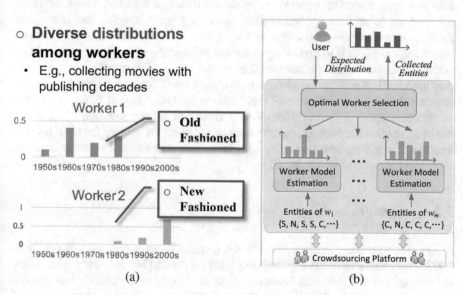

Fig. 7.4 Distribution-aware crowdsourced entity collection. (**a**) Diverse entity distributions. (**b**) Worker selection framework

7.2.2 Crowdsourced Fill

Crowdsourced fill [68] provides workers with a partially filled table and asks the workers to fill the missing values or update the existing wrong answers in the table. To balance the trade-off between cost, latency, and quality, it focuses on *exploiting workers' desire to earn money in order to obtain a table with high quality while without taking too much cost and latency*.

To achieve the goal, Park and Widom [68] develop a system that works as follows: (1) each worker cannot only fill the table but also upvote or downvote the values in a row; (2) requesters can set specific constraints in the values of the table (e.g., collecting any football player with position $=$ "FW" and height$\geq$190 cm); (3) the system can support real-time collaboration among different workers and handle concurrency well; (4) when a table is finalized, the system will allocate the total monetary award to all workers, and the award to each worker is proportional to the contributions made by the worker.

7.3 Crowdsourced Join (Crowdsourced Entity Resolution)

Join is a very important operator in relational database systems. There are many different types of join operators, such as cross join, theta-join, and outer join. Existing crowdsourcing work mainly focuses on equi-join. Given a table (or two tables), an equi-join is to find all *equal* record pairs in the table (or between the two tables). Note that when we say two records are *equal*, it does not only mean that they are identical records but also mean that they are different but refer to the same real-world entity (e.g., "iPhone 4" and "iPhone 4th Gen," or two different pictures of the same person). This problem is also known as entity resolution (ER), which has been studied for decades, but machine-based methods are still far from perfect. Therefore, applying crowdsourcing to solving this problem has recently attracted significant attention [36, 88, 94, 95].

7.3.1 Background

In fact, incorporating humans into the ER process is not a new idea. As early as 2002, a user interactive ER approach has been proposed [74]. However, what's new for crowdsourced ER is that *humans* cannot be simply modeled as a few domain experts, who never make mistakes. Instead, there are hundreds of thousands of ordinary workers (i.e., the crowd) available. They may even provide worse results than machines (without a careful system design). In other words, both humans and machines can make mistakes; thus, it is unclear if we can still combine them together such that the combined one is able to achieve better performance than human-only or machine-only alternatives. In 2012, two research groups [21, 89] answered this question independently by building entity resolution systems on a real-world crowdsourcing platform (i.e., AMT [1]). Both of their systems validated that such goal could be achieved through a good workflow design and quality-control mechanism.

In the following, we will review the recent work on crowdsourced ER. A human-only solution is to ask the crowd to check all n^2 pairs of records. Even with a modest table size (e.g., $n = 100,000$), this method will be very expensive. Hence, most existing works adopt a two-phase framework, where in the first phase, a pruning method is used to generate a candidate set of matching pairs using an inexpensive machine-based technique and in the second phase, the crowd is leveraged to check which candidate pairs are really matching.

7.3.2   Candidate Set Generation

For the candidate set generation phase, a similarity-based method is often adopted [52, 88, 89, 94, 98], which computes a similarity value for each pair of records using a similarity function and takes the pairs whose similarity values are above a threshold as candidate pairs. The method has many advantages when being applied in practice. First, it only needs two input parameters (i.e., a similarity function and a threshold), which does not require a lot of human work for parameter tuning. Second, there are many similarity join algorithms [22, 23, 25, 32, 51–53, 77, 90–93, 104] proposed to efficiently find similar pairs that are above a threshold without enumerating all n^2 pairs. They are some surveys [103] and experimental papers [46]. They are also some systems on Spark and Hadoop for similarity-based processing [22, 80].

If ER tasks become very complex, where a simple similarity-based method cannot generate a very good candidate set, people tend to manually specify domain-specific rules, called blocking rules. For example, the following blocking rule states that if the product names match but the color does not match, then the two products do not match:

$$(\text{name\_match} = Y) \wedge (\text{color\_match} = N) \longrightarrow \text{NO}.$$

Although the use of blocking rules can result in a more satisfactory candidate set, the design of high-quality rules often takes a lot of time. To overcome this limitation, Gokhale et al. [33] propose a crowd-based rule generation approach. Their basic idea is to think a blocking rule as a path in a decision tree, from root to one of the leaf nodes whose label is "No." Based on this idea, they first apply an active learning technique to learn a collection of decision trees from the crowd, and after that, they use the crowd again to check which paths in the decision trees (i.e., candidate blocking rules) make sense.

7.3.3 Candidate Set Verification

In the candidate set verification phase, the goal is to decide which pairs in the candidate set are really matching. Because we need to pay the crowd for doing tasks, existing works explore different perspectives to reduce crowdsourcing cost while still keeping good result quality, e.g., task design [59, 89, 99], leveraging transitive relations [36, 88, 94, 95], and task selection [33, 44, 61, 86, 98].

Task Design Task design mainly involves two problems: user interface (UI) design and task generation.

The single-choice task type is one of the most widely used UIs for crowdsourced ER. It contains a pair of records and asks the crowd to choose "matching" or "non-matching" for the pair. Even for this simple UI, there are many different design choices [59, 99]. For example, we may want to provide a "maybe" option that the crowd can select when they are not sure about their answer, or we can put multiple pairs instead of just a single pair into a task, etc. In addition, the clustering task type has also been used by some works to reduce the cost [59, 89].

Once a particular UI is chosen, the next important question is how to generate tasks for the UI such that all the candidate pairs can be checked. This is a trivial problem for some types of UIs (e.g., single choice) but can be very tricky for others. For example, it has been shown that the cluster-based task generation problem, which aims to generate the minimum number of cluster-based tasks to check a set of candidate pairs, is NP-hard [89].

Leveraging Transitive Relations Leveraging transitive relations is another hot topic for candidate set verification. It is an answer-deduction technique discussed in the previous cost-control chapter. For entity resolution, there are two types of transitive relationships: (1) if $A = B$ and $B = C$, then $A = C$; (2) if $A = B$ and $B \neq C$, then $A \neq C$. After the crowd labels some pairs (e.g., $A = B$, $B \neq C$), we may be able to use the transitive relationships to deduce the labels of some other pairs (e.g., $A \neq C$), thereby saving crowd cost. However, this may not always be true. Consider the same three record pairs. If we label them in a different order (e.g., $A \neq C$ and $B \neq C$), we cannot use transitivity to deduce the label of (A, B), thus still requiring the crowd to label it. Therefore, one natural question is what is the optimal labeling order that can maximize the benefit of using transitivity.

In an ideal case, it is optimal to first present matching pairs to the crowd and then present non-matching pairs [94]. But, this cannot be achieved in practice because it is unknown which pairs are matching or non-matching upfront. Hence, a heuristic approach [94] is proposed to approximate this ideal case, which first computes a similarity value for each candidate pair and then presents the candidate pairs to the crowd in a decreasing order of similarity values. This approach works very well when there exists a good similarity function to compute the similarity values. To handle the case that such similarity function does not exist, some other heuristic approaches have been proposed [88], which can provide a better worst-case performance guarantee than the similarity-based one.

Although transitivity is good for reducing cost, it may have some negative effects on quality and latency. In terms of quality, transitivity will amplify the impact of crowd errors on ER results. For example, suppose $A = B$ and $B = C$. If the crowd falsely label them as $A = B$ and $B \neq C$, such error will be propagated through transitivity, resulting in an erroneous label of $A \neq C$. Some ideas such as using correlation clustering [95] or designing new decision functions [36] have been explored to tackle this issue. In terms of latency, in order to leverage transitivity, we cannot present all candidate pairs to the crowd at the same time. Instead, it should be an iterative process, where only a single pair or a few pairs can be presented to the crowd at each iteration. As a result, the iterative process may take much longer time to complete because it does not fully utilize all available crowd workers. This is considered as a main challenge in the use of transitivity for crowdsourced ER, and various parallel algorithms are proposed to accelerate the process [94, 95].

Task Selection Because the crowd is costly, sometimes it may be infeasible to ask them to check all candidate pairs. When there is only a limited amount of monetary budget for crowdsourcing, a natural idea is to explore how to select most valuable candidate pairs for the crowd to check. Existing works on this topic have different selection objectives: query-workload driven [44], ER-result driven [86, 98], and classifier driven [33, 61].

A query-workload driven approach assumes the existence of a query workload, and it aims to select those pairs such that when they are checked by the crowd, it will have the most benefit for the query workload. Inspired by the concept of the value of perfect information (VPI), Jeffrey et al. [44] develop a decision-theoretic framework in data-space systems to achieve this goal.

Unlike the workload-driven approach, an ER-result-driven approach has a different optimization goal. It aims to select those pairs that are most valuable to the ER-result quality. The approach often models candidate pairs as a probabilistic graph, where each edge in the graph represents a candidate pair and the probability of each edge represents the probability that the corresponding candidate pair is a matching pair. The probabilistic graph is used to measure the quality of the current ER results and then predicate which pairs should be selected that can maximize the improvement of the quality [86, 98].

A classifier-driven approach uses the crowd to train a classifier. Its goal is to select the candidate pairs that are most beneficial to the training process of the classifier. According to different characteristics of the classifiers, they have different selection strategies. The random forest classifier selects the pairs based on the percentage of decision trees that have contradictory classification results [33]; the SVM classifier uses the distance of each pair to the decision hyperplane to decide which pairs to choose [61].

7.3.4 Human Interface for Join

In order to interact with workers in crowdsourcing, the requester needs to design friendly interfaces. For tasks about crowd-based join, there are mainly two kinds of interfaces. One is *pairwise interface* and the other is *multi-item interface*. For example, suppose there are a number of names of products to be matched. One can ask workers to compare every pair of them, or design an interface like Fig. 7.5, which combines k items in one interface and asks workers to label them. The disadvantage of pairwise interface is that it generates a large number of questions, which involves huge cost. But the multi-item interface reduces the cost by grouping some items together. However, the multi-item interface also has a disadvantage. It may result in low quality, especially for difficult problems because workers may lose concentration and patience faced with many pairs to be compared. But the pairwise interface achieves high quality because workers only focus on one pair at each task. Considering the advantages and disadvantages of the two interfaces, Vasilis et al. [87] proposed Waldo, a framework that uses the two interfaces to reduce cost while keeping high quality.

The Waldo framework is applied in a loop. In each iteration, it selects tasks to ask workers based on initial machine pruning results and workers' answers from previous iterations. At the beginning of each iteration, they classify the pairs into resolved pairs and unresolved pairs, which is decided by whether their probabilities of being matched are above a threshold or not. In addition, they also apply an entity resolution algorithm to help some unresolved pairs to be resolved using transitivity. For these unresolved pairs, Waldo takes them as input of the "Detect Difficult Pairs" stage. The output of this is a classification of the unresolved pairs into difficult and non-difficult pairs. For difficult pairs, they tend to use pairwise interfaces to display them to achieve high quality; while for non-difficult pairs, they tend to use multi-item interfaces to reduce cost. To address this problem, they propose an effective optimal grouping approach. After choosing the interfaces, they publish the tasks to the crowdsourcing platform and obtain their answers. Waldo repeats this until all pairs are resolved. The experimental results show that Waldo outperforms the existing methods in terms of both cost and quality.

Fig. 7.5 Multi-item interface

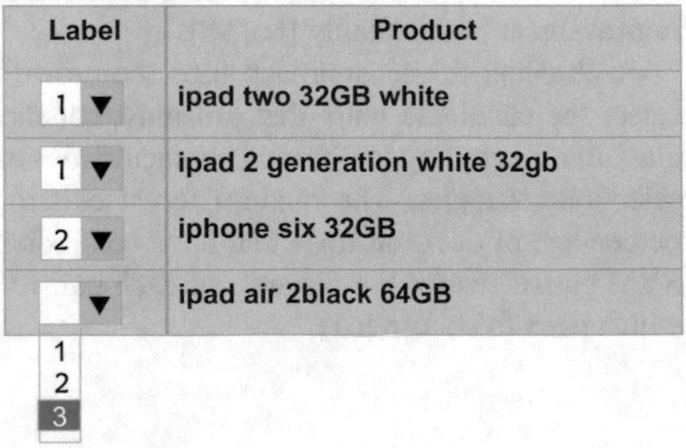

Label	Product
1 ▼	ipad two 32GB white
1 ▼	ipad 2 generation white 32gb
2 ▼	iphone six 32GB
▼	ipad air 2black 64GB

1
2
3

7.3.5 Other Approaches

Partial-Order Approach Chai et al. [15] proposed a partial-order approach to reduce the monetary cost of crowdsourced joins with little loss in quality. The basic idea is to first define a partial order over the record pairs and then prune many pairs that do not need to be asked based on the partial order. Specifically, a partial order is as follows: (1) if a pair of records refer to the same entity, then the pairs preceding this pair also refer to the same entity; (2) if a pair of records refer to different entities, then the pairs succeeding this pair refer to different entities. In particular, given two pairs $p_{ij} = (r_i, r_j)$ and $p_{i'j'} = (r_{i'}, r_{j'})$, if (r_i, r_j) has no smaller similarity than $(r_{i'}, r_{j'})$ on every attribute, then we denote this by $p_{ij} \succ p_{i'j'}$, which means that p_{ij} precedes $p_{i'j'}$. The method selects a pair as a question and asks the crowd to check whether the pair of records refer to the same entity. Based on the answer of this pair, they infer the answers of other pairs based on the partial order. For example, in Fig. 7.6 [15], they take each node as a pair of records, and these green (red) ones represent matching (non-matching) pairs. According to the partial-order approach, if the requester asks p_{45} and workers say that it is a matching pair, then they color p_{35}, p_{36}, p_{46}, p_{12}, and p_{34} as green without asking the crowd to check them. If the requester asks p_{14} and workers say that it is a non-matching pair, then they color p_{13} and p_{26} as red. Thus, the goal is to judiciously select the pairs to ask in order to minimize the number of asked pairs. To this end, they devise effective algorithms to iteratively select pairs until they get the answers of all the pairs. To further reduce the cost, they propose a grouping technique to group the pairs such that they only need to ask one pair instead of all pairs in each group. Since asking only one pair in each iteration leads to high latency, they propose effective techniques to select multiple pairs in each iteration. As both the partial order and the crowd

Fig. 7.6 An example of the partial-order approach

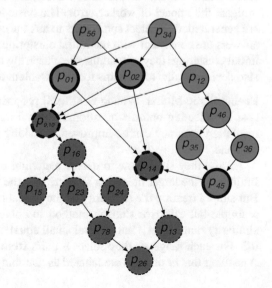

may introduce errors, they develop error-tolerant techniques to tolerate the errors. The basic idea is that given a pair, if its answers from multiple workers have high uncertainty (e.g., two workers give Yes and three workers give No), they do not use this pair to infer its ancestors' or descendants' answers. Instead, they put the pairs into histograms based on their similarities (pairs with close similarities are put into the same histogram) and then infer the results based on the majority voting of pairs in the histogram. Zhuang et al. [108] extended this approach for entity matching in large-scale knowledge bases.

Correlation Clustering Approach Transitivity is a commonly used technique to reduce the cost, but it may result in low accuracy due to worker errors. For example, given two sets of records A and B where the records in each set have been identified by workers as the same entity. Now one would like to judge the relationships between the records of the two sets. Using the transitivity, we just need to pick up one record from each set, respectively, and then ask the crowd to check whether they refer to the same entity or not. However, if workers give a wrong answer about the pair, the error will be amplified due to the use of transitivity. To address this issue, Wang et al. [95] proposed a correlation clustering-based framework, which mainly consists of three steps: a pruning phase, a cluster generation phase, and a cluster refinement phase.

Firstly, given a set of records, the pruning phase applies a machine-based algorithm to identify some dissimilar pairs as non-matching pairs. Then, the cluster generation phase clusters the remaining pairs using correlation clustering. The main idea is that every pair has a similarity score, which is computed through asking multiple people. For example, if four out of five people say one pair is a matching pair, the score of this pair is 0.8. Then the correlation clustering tries to make the pairs in the same cluster as similar as possible while the pairs between different clusters as dissimilar as possible. Finally, the cluster refinement phase is used to mitigate the impact of worker error. The basic idea is that after the initial clusters are generated, they select some pairs to ask the crowd and check whether the crowd answers are consistent with the initial clustering. If many answers conflict with the initial clustering, they will adjust the clustering to verify the errors. Moreover, they also devise parallel algorithms for this problem to reduce latency.

Probabilistic Model Whang et al. [98] proposed a probabilistic model to reduce cost by selecting optimal questions. Their framework evaluates how much entity resolution accuracy can be improved by asking each question and then selects the best one.

Firstly, they study how to derive matching probability from string similarities. Intuitively, the larger the string similarity is, the higher the matching probability is. But simply treating the similarity as the probability is not accurate. They proposed a simple but effective statistic method to solve the problem. They first split the similarity range ([0,1]) into several small equal ones, such as five ranges with width 0.2. For each range r_i, they select n pairs from it and ask workers to label them. Assuming that m of them are labeled as matching pairs, they consider that the pairs from this range will have a matching probability of $\frac{m}{n}$. For example, in Fig. 7.7, since pairs whose similarity scores lie in [0.8,1] are all matching pairs, they infer that the matching probability of these pairs is 1. Therefore, in Fig. 7.8, given the similarities of the three pairs on the left side, we can obtain their matching probabilities on the right one.

Once matching probabilities are obtained, they use *possible worlds* to select the best question. A possible world represents one possibility of workers' answers. For example, the first graph in Fig. 7.9 means that workers consider all three pairs as matching. The probability of this possible world is $1 \times 0.3 \times 0.7 = 0.21$. And the possible worlds S can be easily deduced by C using transitivity. For example, for the first three possible worlds in C, they all lead to the conclusion that a, b,

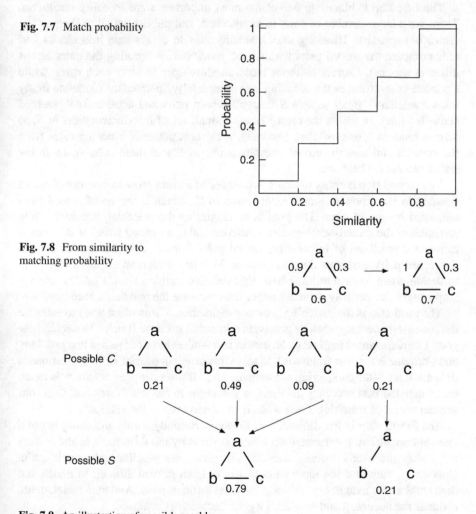

Fig. 7.7 Match probability

Fig. 7.8 From similarity to matching probability

Fig. 7.9 An illustration of possible worlds

and c are matching. Therefore, the first possible world in S uses the sum of their probabilities. The authors assume that if workers say that a pair is matching, its matching probability becomes 1. Using the possible worlds, they can compute the expected accuracy for each question and then select the question that maximizes the expected accuracy.

Hands-Off Approach Although the existing methods above leverage the crowd's ability to improve the quality of entity resolution, they still need to get developers involved. Therefore, it is hard for some non-expert enterprises and crowdsourcing startups and masses to use. To address this problem, Gokhale et al. [33] proposed Corleone, a hands-off approach for entity matching. This approach can use the crowd to do all four major steps in entity resolution.

The first step is blocking, one of the most important steps in entity resolution. There are a large number of pairs to be matched, and publishing all of them to the crowd is expensive. Blocking uses heuristic rules to divide data into blocks and only compare the record pairs within each block (while ignoring the pairs across different groups). Current methods require a developer to write such rules, while Corleone crowdsources the blocking step completely. Specifically, Corleone firstly takes a relatively small sample S from all record pairs and applies crowdsourced active learning, in which the crowd labels a small set of informative pairs in S, to learn a random forest matcher. Secondly, it extracts potential blocking rules from the matcher and uses the crowd to evaluate the quality of these rules again. In the end, it retains the best ones.

The second step is entity matcher. A matcher M applies crowdsourcing and active learning to learn how to match record pairs in C, which is the set of record pairs outputted by the blocker. The goal is to maximize the matching accuracy while minimizing the crowdsourcing cost. Corleone trains an initial matcher M, uses it to select a small set of informative record pairs from C, asks the crowd to label the record pairs, and uses them to improve M. They design an effective algorithm to decide when to stop training because excessive training wastes money and yet surprisingly can actually decrease rather than increase the matcher's accuracy.

The third step is the matching accuracy estimation. Users often want to estimate the matching accuracy such as precision and recall because it helps to decide how good a crowdsourced method is. However, few studies have addressed this problem, and Corleone is the first framework to solve this problem deeply. Corleone estimates M's accuracy after completing the matcher step. If the estimated accuracy is much better than the best accuracy obtained, it will come to the fourth step and then start another round of matching. Otherwise, it stops and returns the results.

The fourth step is the difficult pair detector. Actually, entity matching is not a one-shot operation. If estimated accuracy obtained by the third module above does not satisfy the user's request, they have to revise the pipeline and match again. Concretely, they find out tuple pairs that have been proven difficult to match and then build a new matcher specifically for these difficult pairs. And then match them, estimate the accuracy, and stop until a good accuracy is achieved.

7.4 Crowdsourced Sort, Top-k, and Max/Min

Definition 7.7 (Crowdsourced Sort, Top-k and Max/Min) Given a set of items $\mathcal{O} = \{o_1, o_2, \cdots, o_n\}$, where the items are comparable but hard to compare by machines (e.g., comparing the clarity of images), crowdsourced top-k aims to find a k-size item set $\mathcal{R} = \{o_1, o_2, \cdots, o_k\}$ where o_i is preferred to o_j (denoted by $o_i \succ o_j$) for $o_i \in \mathcal{R}$ and $o_j \in \mathcal{O} - \mathcal{R}$ (finding max/min is a special case of top-k); and crowdsourced sort aims to sort the items and gets a sorted list $o_1 \succ o_2 \succ \cdots \succ o_n$.

Zhang et al. [107] give an experimental survey on this problem.

Evaluation Metrics Crowdsourced sort aims to get the correct comparison answer for every pair. If an algorithm returns an incorrect pair, i.e., it returns $o_i \succ o_j$ but actually $o_j \succ o_i$, the algorithm involves an error. Thus, the number of incorrect pairs is used to evaluate the quality of a crowdsourced sort algorithm.

For crowdsourced top-k, there are two widely used metrics. The first is recall, which is the ratio of correctly returned top-k results to k. The second is a combination of recall and the number of incorrect pairs by considering the order of the top-k results.

7.4.1 Workflow

To utilize the crowd to find top-k items, we need to generate crowdsourced tasks. There are two widely used ways to generate crowdsourced tasks. The first is single choice, which selects two items and asks the crowd to select the preferred one (a.k.a., pairwise comparison). The second is rating, which selects multiple items and asks the crowd to assign a rate to each item. The rating-based method has some weaknesses. First, the crowd prefers pairwise comparisons than ratings as the former is much easier. Second, it is rather hard for the crowd to assign an accurate rate, and items in different rating groups may not be correctly compared. Thus, the rating-based method usually has a lower accuracy than pairwise comparison [50, 59]. Most of existing works use pairwise comparisons. Next we introduce the pairwise comparison-based framework. We also introduce some hybrid methods that use both rating and comparison questions later.

7.4.2 Pairwise Comparisons

Pairwise comparison methods use the single-choice task type, where each task is to compare two items. To reduce the monetary cost, existing methods employ a task selection strategy, where b pairs are crowdsourced in each round. Based on the comparison answers of the crowdsourced pairs, it decides how to select b pairs in the

next round. To reduce crowd errors, each task is assigned to multiple workers, and
the final result is aggregated based on the answers of these workers, e.g., weighted
majority vote.

Graph Model To model the results, *a directed graph* is constructed where nodes
are items and edges are aggregated comparison answers. For each pair (o_i, o_j), if the
aggregated result is $o_i \succ o_j$, there is a directed edge from o_i to o_j where the weight
is the aggregated preference, e.g., the percentage of workers that select $o_i \succ o_j$
among all the workers that are asked to compare the pair.

7.4.3 Result Inference

Given the answers for the generated tasks, the result inference tries to infer the query
results, i.e., top-k items or a sorted list. There are five types of methods: (1) score-
based methods, (2) iterative reduction methods, (3) machine learning methods, (4)
heap-based methods, and (5) hybrid methods.

(1) Score-Based Methods

Guo et al. [37] prove that finding the top-1 item (i.e., finding the max/min) with
the largest probability is NP-hard by a reduction from Kemeny rankings. Thus,
inferring the top-k items is also NP-hard, and some score-based algorithms are
proposed, which assign each item o_i with a score s_i and select the k items with the
largest scores as top-k results (or sort them based on the scores). Next we discuss
how to assign scores.

BordaCount [2] The score of o_i is its out-degree $d^+(o_i)$ (the number of wins
compared with its out-neighbors), i.e.,

$$s_i = d^+(o_i).$$

Copeland [70] The score of object o_i is its out-degree $d^+(o_i)$ minus its in-degree
$d^-(o_i)$ (the number of wins minus the number of losses), i.e.,

$$s_i = d^+(o_i) - d^-(o_i).$$

Local [37] The above two methods only consider the neighbors of each item and
cannot capture more information. Local top-k algorithm [37] is proposed to address
this problem by considering two-hop neighbors. Obviously if an item has more two-
hop out-neighbors (i.e., its out-neighbors' out-neighbors), the item will beat more
items (based on transitivity), and thus the item has a larger score. Similarly, if an
item has more two-hop in-neighbors (i.e., its in-neighbors' in-neighbors), the item
will be beaten by more items, and thus the item has a lower score. Accordingly, the
overall score of an object is computed as below:

$$s_i = M_{i*} - M_{*i} + \sum_{o_j \in \mathcal{D}^+(o_i)} M_{j*} - \sum_{o_j \in \mathcal{D}^-(o_i)} M_{*j},$$

where M_{ij} (M_{ji}) is the number of workers who prefer o_i to o_j (who prefer o_j to o_i). $M_{i*} = \sum_j M_{ij}$, $M_{*i} = \sum_j M_{ji}$, and $\mathcal{D}^+(o_i)$ is the out-neighbor set of o_i and $\mathcal{D}^-(o_i)$ is the in-neighbor set of o_i.

Indegree [37] It computes the score based on the Bayesian model. It first computes the probability of $o_i \succ o_j$ given the aggregated preference and then uses the probability to compute the score for each object. The probability of $o_i \succ o_j$ given M_{ij} and M_{ji} is

$$P(o_i \succ o_j | M_{ij}, M_{ji}) = \frac{P(M_{ij}, M_{ji} | o_i \succ o_j) P(o_i \succ o_j)}{P(M_{ij}, M_{ji})}$$

$$= \frac{P(M_{ij}, M_{ji} | o_i \succ o_j)}{P(M_{ij}, M_{ji} | o_i \succ o_j) + P(M_{ij}, M_{ji} | o_j \succ o_i)},$$

$$P(M_{ij}, M_{ji} | o_i \succ o_j) = \binom{M_{ij} + M_{ji}}{M_{ij}} p^{M_{ij}} (1 - p)^{M_{ji}};$$

$$P(M_{ij}, M_{ji} | o_j \succ o_i) = \binom{M_{ij} + M_{ji}}{M_{ji}} p^{M_{ji}} (1 - p)^{M_{ij}}.$$

p is the estimated worker accuracy which is a fixed value. Besides, it assumes $P(o_i \succ o_j) = P(o_i \prec o_j)$. Then, it computes the score of object o_i as below:

$$s_i = \sum_{j \neq i} P(o_i \succ o_j | M_{ij}, M_{ji}).$$

Modified PageRank(MPageRank) [37] It extends the original PageRank by considering the crowdsourced comparisons and computes the score of each item:

$$s_i = pr_k[i] = \frac{1 - c}{n} + c \sum_{i \neq j} \frac{M_{ij}}{M_{*j}} pr_{k-1}[j],$$

RankCentrality [63] The algorithm adopts an iterative method based on random walk to compute the score of each object. Its core component is to construct an $n \times n$ transition matrix P, in which $P_{ij} = \begin{cases} \frac{1}{d_{max}} w_{ji} & \text{if } i \neq j \\ 1 - \frac{1}{d_{max}} \sum_{k \neq i} w_{ki} & \text{if } i = j \end{cases}$, where w_{ji} is the weight of edge from o_j to o_i and d_{max} is the maximum in-degree of a node (i.e., $d_{max} = \max_{o_i} d^-(o_i)$). Then, it initializes a $1 \times n$ matrix $\hat{s}$ and computes $\hat{s} \times P \times P \times \cdots$ until convergence. The score of o_i is computed as

$$s_i = \hat{s}[i] \text{ where } \hat{s} = \lim_{t \to \infty} \hat{s} \times P^t.$$

ELO [26] It is a chess ranking system and can be used to compute the score s_i. The basic idea is that, if item o_i with higher ranking beats another lower one o_j, only a few scores will be added to s_i; on the contrary, if o_j wins, a lot of scores will be added to s_j. Formally, each object o_i is randomly assigned a score s_i initially. When o_i is compared with o_j, the two scores s_i and s_j will be updated as below:

$$s_i = s_i + H \left(C_i - \frac{1}{1 + 10^{(s_j - s_i)/400}} \right);$$

$$s_j = s_j + H \left(1 - C_i - \frac{1}{1 + 10^{(s_i - s_j)/400}} \right),$$

where H is a tuning parameter (set to 32 by default). $C_i = 1$ if $o_i \succ o_j$; $C_i = 0$ otherwise.

Balanced Rank Estimation(BRE)/Unbalanced Rank Estimation(URE) [96]
The score is computed based on the probability theory. The balanced rank estimation (BRE) considers both incoming and outgoing edges. To compute the score s_j, it computes the relative difference of the number of objects proceeding and succeeding o_j:

$$s_j = \frac{\sum_{i \neq j} b_{ij}(2w_{ij} - 1)}{2\alpha n} \propto \sum_{i \neq j} b_{ij}(2w_{ij} - 1).$$

where α is the selection rate (i.e., $\alpha\binom{n}{2}$ pairs will be compared); $w_{ij} = 1$ if $o_i \succ o_j$; $w_{ij} = 0$ otherwise; and $b_{ij} = 1$ if o_i and o_j are compared by the crowd; $b_{ij} = 0$ otherwise.

The unbalanced rank estimation (URE) computes the score of o_i only based on its incoming edges. To compute the score s_j, it computes the number of objects proceeding o_j:

$$s_j = \frac{1}{\alpha n} \sum_{i \neq j} b_{ij} w_{ij}^n \propto \sum_{i \neq j} b_{ij} w_{ij}^n,$$

(2) Iterative Reduction Methods

The iterative-reduce methods adaptively eliminate the low rank items that have small possibilities in the top-k results, until k items left.

Iterative [37] It first utilizes the score-based methods to compute the scores of each item and then removes a half of items with the smallest scores. Next, it recomputes the scores on the survived items and repeats the iterations until k items left.

PathRank [27] The main idea of PathRank is to perform a "reverse" depth first search (DFS) for each node, which traverses the graph by visiting the in-neighbors of each node. If it finds a path with length larger than k, it eliminates the item as k items have already been better than the item.

AdaptiveReduce [27] Initially there are n items. Then it selects an informative set and utilizes the set to eliminate items with small possibilities in the top-k answers. It repeats this step using the survived items until finding top-k items.

Bound-Based Methods [14] The bound-based methods build a stochastic matrix and use several algorithms to identify the top-k results based on the matrix: (i) *Sampling Strategy with Copeland's Ranking (SSCO)*. It selects the top-k rows with most entries above 0.5 as the top-k results. (ii) *Sampling Strategy with Sum of Expectations (SSSE)*. It selects the top-k rows with the largest average value as the top-k results. (iii) *Sampling Strategy based on Random Walk (SSRW)*. It first computes a stochastic matrix and then derives its principal eigenvectors (that belong to the eigenvalue 1). Then it identifies the top-k rows with the largest eigenvalues as the top-k answers.

(3) Machine Learning Methods

These methods assume that each item has a latent score which follows a certain distribution. Then they utilize machine learning techniques to estimate the score. Finally, they use the latent scores to sort the items or get top-k items.

CrowdBT with Bradley-Terry Model [13] The Bradley-Terry (BT) model can be used to estimate the latent score [13]. In the BT model, the probability of $o_i \succ o_j$ is assumed as $\frac{e^{s_i}}{e^{s_i}+e^{s_j}}$. Then based on the crowdsourced comparisons, it computes the latent scores by maximizing $\sum_{o_i \succ o_j \in \mathcal{L}} \log(\frac{e^{s_i}}{e^{s_i}+e^{s_j}})$, where $\mathcal{L}$ is a set of crowdsourced comparison answers. But the BT model does not consider the workers' qualities. To address it, Chen et al. [16] propose the CrowdBT model, assuming that each worker has a quality η_w as discussed in Worker Probability (see Chap. 3).

CrowdGauss with Gaussian Model [69] It assumes that the score follows the Gaussian distribution, where the score is the mean of the distribution. The probability of $o_i \succ o_j$, i.e., $P(o_i \succ o_j)$, can be computed by the cumulative distribution function (Φ) of the two standard Gaussian distributions, i.e., $P(o_i \succ o_j) = \Phi(s_i - s_j)$. Then CrowdGauss computes the scores by maximizing $\sum_{o_i \succ o_j \in \mathcal{L}} M_{ij} \cdot \log(\Phi(s_i - s_j))$ where M_{ij} is the number of workers reporting $o_i \succ o_j$.

HodgeRank [45] In order to estimate a global order for n items, HodgeRank [45] utilizes a matrix decomposition-based techniques to compute the score.

TrueSkill [40] TrueSkill improves ELO by reducing the repeated times as ELO needs to repeat many times to convergence. Different from ELO, the score of each item o_i is represented by a Gaussian distribution $N(s_i, \delta_i)$, where s_i is the estimated score for o_i and δ_i is the deviation of s_i. For each crowdsourced answer $o_i \succ o_j$, it updates the scores and deviations.

(4) Heap-Based Methods

Two-Stage Heap [20] In the first phase, the items are divided into $\frac{n}{X}$ buckets (where $X = \frac{xn}{k^2}$) such that the probability of two top-k items appearing in the same bucket is at most x. In each bucket, a tournament-based max algorithm [20] is conducted to select the best item in the bucket. Each pair on top levels of the tournament is compared multiple times, and each pair on low levels of the tournament is compared only once. The second phase utilizes a heap-based method [31] to identify the top-k results from these best items. To tolerate errors, when constructing and re-heapifying the heap, each pair is compared by multiple workers, and the algorithm uses the majority voting to obtain a combined preference. After popping an item from the heap, the algorithm asks next pairs following the re-heapifying order.

(5) Hybrid Methods

There are three algorithms [50, 54, 101] that combine rating and comparison tasks. They utilize the rating and comparison answers to learn the score. For rating, they predefine τ categories and each category has a range. If the score of item o_i falls in a range, o_i belongs to the corresponding category.

Score Combination [101] It first selects some rating and comparison tasks and then infers the scores based on these results. The score for each item o_i is modeled by $s_i + \varepsilon_i$, where $\varepsilon_i \sim N(0, \delta^2)$, which is utilized to tolerate crowd errors. For rating, it computes the probability of o_i being in the category χ_c by standard Gaussian distribution. For comparison, it constructs the comparison matrix M and computes the probability of observing M based on the Bayesian theory. Then it combines rating and comparison to infer the results.

Rating First [50] It first crowdsources all rating tasks and then selects some candidate items with higher ratings (prunes those with low ratings to reduce the number of comparison pairs). Next it chooses some pairs from the candidates as comparison tasks. The score of each item o_i is modeled as a normal variable $N(s_i, \delta^2)$. Given the rating results E_R and comparison answers E_C, it assumes that these results are gotten independently and computes the scores based on the results. However, the maximum likelihood estimation is rather expensive; thus, a modified PageRank approximation [50] is proposed to estimate the score for each item.

Adaptive Combine [54] It contains two main steps. The first step infers top-k results based on the current answers of rating and ranking questions, called *top-k inference*. It models the score of each object as a Gaussian distribution, utilizes the rating and ranking results to estimate the Gaussian distribution, and infers the top-k results based on the distributions. The second step selects questions for a coming worker, called *question selection*. Based on the probability of an object in the top-k results, it can get two distributions: real top-k distribution and estimated top-k distribution. Thus, it proposes an effective question-selection strategy that selects questions to minimize the distance between the real distribution and the estimated distribution. As it is rather expensive to minimize the difference, it proposes effective heuristics to improve the performance.

7.4.4 Task Selection

The task selection decides which candidate pair of items should be asked next. There are three types of methods: (1) heuristic-based methods, (2) bound-based methods, and (3) active learning methods.

(1) Heuristic-Based Methods

Guo et al. [37] prove that selecting the pairs to maximize the probability of obtaining the top-k results given the comparison answers of arbitrary crowdsourced pairs is NP-hard and propose four heuristics, which are designed for selecting the max (top-1) result. The algorithms first compute a score s_i for item o_i as discussed above. Suppose the sorted items based on the scores are $o_1, o_2, \cdots, o_n$. Then, the algorithms select the next b pairs as follows.

Max It selects b pairs: $(o_1, o_2), (o_1, o_3), \cdots, (o_1, o_{b+1})$.

Group It groups the i-th item with the $(i+1)$-th item and selects $(o_1, o_2), (o_3, o_4)$, $\cdots, (o_{2b-1}, o_{2b})$.

Greedy It selects the pairs based on $s_i \times s_j$ in descending order and selects b pairs with the largest value.

Complete It first selects x items with the highest scores, where x is the largest number satisfying $\frac{x*(x+1)}{2} \leq b$. The selected pairs include two parts. The first part includes all $\binom{x}{2}$ pairs among these x items. The second part contains $(o_1, o_{x+1}), (o_2, o_{x+1}), \cdots, (o_{b-\frac{x*(x+1)}{2}}, o_{x+1})$.

(2) Bound-Based Methods

SSCO and SSSE [14] estimate a bound for each pair and utilize the bound to select next pairs. They first compute a confidence interval $[l_{ij}, u_{ij}]$, where l_{ij} (u_{ij}) is the lower (upper) bound of the probability of $o_i \succ o_j$. Based on the confidence interval, they select a set S and discard a set D. Since the relationships between pairs in $S \cup D$ have been effectively captured, these pairs do not need to be compared. Thus, they select pairs that are not in $S \cup D$. In comparison, SSRW [14] selects the next pairs by random walk on the stochastic matrix.

(3) Active Learning Methods

CrowdGauss [69] The scores can be modeled by a multivariate Gaussian distribution $N(\hat{s}, C)$, where $\hat{s}$ is a $1 \times n$ matrix indicating the score for all the items (initialized by random values) and C is the covariance matrix of $\hat{s}$ (C_{ij} is the value at

i-th row and j-th column). In each round of pair selection, the expected information gain for each pair (o_i, o_j) is computed, and it selects the pair with the largest expected information gain and updates $\hat{s}$ and C.

CrowdBT [16] The above method does not consider the worker quality. Chen et al. [16] propose an active learning method by taking into account the worker quality. The score of item o_i is modeled by a Gaussian distribution $N(s_i, \delta_i)$, and the quality of worker w is modeled by Worker Probability (see Chap. 3).

Combine [101] Instead of just utilizing pairwise comparisons, Ye et al. [101] propose an active learning strategy by combining rating and comparison together. Given a budget, it selects some rating-based questions and comparison-based questions to maximize the expected information gain.

Distribution-Aware [54] It first computes the top-k probability distribution for ground truth D_G and the estimated probability distribution D_R, and then the goal of question selection is to narrow the distance between D_G and D_R as soon as possible by utilizing both rating and ranking questions. The distance is evaluated by the relative entropy of D_G and D_R, which is $KL(D_G||D_R)$. For rating questions, it enumerates every object and computes the expected improvement for the probability curve, respectively. For ranking questions, assuming the size of the ranking question is y, there are total $\binom{n}{y}$ combinations of objects. It calculates the expected improvement of the probability curve for each combination. Then it selects the question to minimize the distance $KL(D_G||D_R)$ from those n rating questions and $\binom{n}{y}$ ranking questions. If multiple questions should be selected, it selects questions based on the distance in descending order. It also develops effective algorithms to estimate the probability distribution and devises efficient question-selection algorithms.

7.4.5 Crowdsourced Max

Crowdsourced max studies how to use the crowd to find the *max* item in a dataset, i.e., $k = 1$ for the top-k problem. The item comparison is often based on human subjectivity, such as finding the most beautiful picture about Golden Gate Bridge and returning the best Chinese restaurant in San Francisco. Besides the methods for top-k, there are two types of solutions that are specifically designed to crowdsourced max: structured [85] and unstructured [37].

The structured solution generates tasks in a predefined structure. It is guaranteed that after tasks are finished using the structure, the max item can be directly deduced based on transitivity (i.e., if A is better than B and B is better than C, then A is better than C). Venetis et al. [85] propose two max algorithms based on this idea: (1) The bubble max algorithm is derived from the well-known bubble sort algorithm. It first asks the crowd to compare (A, B). Suppose A is better than B. Then, it asks the crowd to compare the better one (i.e., A) with C. Similarly, after the crowd compares all $n - 1$ pairs, the final item will be considered as the best one. (2) The tournament max algorithm is a tournament-like algorithm that is widely used in sports for choosing the best team. It first generates $\frac{n}{2}$ pairs, i.e., (A, B), (C, D), (E, F), etc. and asks the crowd to compare them. Then, it chooses the better one of each pair and generates $\frac{n}{4}$ new pairs. Iteratively, when there is only one item left, it will be considered as the best one. It has been shown that the tournament max algorithm performs better than the bubble max algorithm. To further improve the performance, Venetis et al. also develop a hill-climbing-based approach to adaptively tune two parameters used in the max algorithms.

Unlike the structured solution, the unstructured solution generates tasks in a dynamic way. It utilizes the results of finished tasks to dynamically decide which tasks should be generated next. Guo et al. [37] identify two main problems (result inference and task selection) in this setting and formalize them based on maximizing the likelihood function.

7.5 Crowdsourced Aggregation

Aggregation queries (e.g., count and median) are widely used in data analysis. This section will first describe how to use the crowd to deal with aggregation functions and then discuss how to support the group by clause.

7.5.1 Crowdsourced Count

Crowdsourced count studies how to use the crowd to count the number of items in a dataset that satisfy some given constraints. At first glance, this is quite similar to crowdsourced filtering. But, the difference is that the crowdsourced count does not require returning the satisfied items but only a number. Thus, there are some interesting new ideas (Fig. 7.10).

The first idea is to use sampling [58]. Instead of checking all the items in the entire dataset, this idea only checks the items in a random sample. Sample estimation is a well-studied topic in statistics.

Definition 7.8 (Sample Estimation) Let $|D|$ be the data size, $|S|$ be the sample size, and k be the number of the items in the sample that satisfies the constraints. The count w.r.t the full data can be estimated from the sample as $k \cdot \frac{|D|}{|S|}$. This is an unbiased estimate, and the error of the estimate decreases proportional to $\frac{1}{\sqrt{|S|}}$.

In addition to sampling, another interesting idea is a new user interface, called *count-based interface* [58]. Recall that crowdsourced filtering mainly uses a *single-choice interface*, which asks the crowd to choose "Yes" or "No" for an item, indicating whether the item satisfies the constraints. In comparison, a count-based

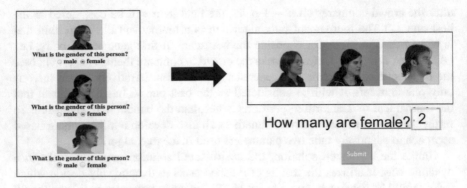

Fig. 7.10 Interface design for crowdsourced count

interface shows a small batch of items (e.g., ten photos) to the crowd and asks them to approximately estimate the number of items that satisfy the constraints, e.g., counting how many of the ten photos contains mountains. Interestingly, the two interfaces are suitable for different situations. The count-based interface is more suitable for images, but the single-choice interface performs better for text. This might be because humans are better at processing batches of images than strings of text.

A slight variant of the counting problem is to count the number of objects in a *single* photo (e.g., how many persons are there in one photo?). In this problem, a new challenge is how to split the photo into small segments such that each segment only contains a small number of objects. The reason for doing that is that humans are often good at batch counting, but the batch size cannot be very large. For example, showing a photo with hundreds of people to a worker to count will lead to long waiting time and low-quality answers. To address this problem, [75] explores two scenarios. One assumes that there is no good computer vision approach available that can help to identify the objects in the photo and a top-down algorithm is proposed to generate image segments; the other is a bottom-up approach which first uses a computer vision approach to identify the objects in the photo and merges the image segments containing an identified object into bigger ones.

7.5.2   Crowdsourced Median

A median operator aims to find the centroid in a set of items. It is a key operation in various clustering algorithms (e.g., k-median). One simple crowdsourced implementation is to apply a crowdsourced sort operator to the set of the items and then return the middle item in the sorted list. Because the median operator only needs to return the centroid, the expensive sorting process can actually be avoided using the crowd-median algorithm [39]. The algorithm presents a task with three items to the crowd and asks them to pick the item (called "outlier") that is different from the other two. If the underlying data follows a univariate normal distribution, the centroid often has the highest probability of being an outlier. Thus, the problem is reduced to finding the item with the highest probability. Note that the algorithm does not enumerate all $\binom{n}{3}$ possible tasks to compute the probabilities, and instead it uses sampling to estimate the probability for each item.

7.5.3 Crowdsourced Group By

A crowdsourced group-by operator can group a set of items based on their unknown type. It is often used with aggregation functions (e.g., COUNT, MAX) to enable more advanced data analysis. For example, the following group-by aggregation query

```
SELECT BEST-LOOKING(photo) FROM table
GROUP BY PERSON(photo)
```

will find the best looking photo of each individual person. In the database community, prior work [20] assumes that there is an underlying ground truth for the type of each item. For example, the ground truth of the above query is a person's name. Based on this assumption, they can derive good theoretical bounds on the minimum number of tasks that are required to ask for crowds. In the machine learning community, there are some works that do not need to make this assumption. Their basic idea is to treat group-by as a clustering problem and to study how to incorporate crowdsourcing into existing clustering algorithms [34, 39, 102].

7.6 Crowdsourced Categorization

Given a set of categories and a set of uncategorized (i.e., unlabeled) objects, the object categorization problem aims to ask the crowd to find the most suitable category (i.e., label) for each object [67]. Each task is a single-choice question – given an object and a category, each task asks whether the object belongs to the category or not. For example, given a photo of "Toyota Corolla" and a category hierarchy (e.g., vehicles, cars, Toyota, Corolla), it aims to categorize the photo into the most appropriate category, e.g., asking whether the photo is in category of "Toyota."

There are three dimensions that characterize the different instances of the category problem.

Definition 7.9 (Single or Multiple Target Categories) In the single case, each object has a single target category; while in the multiple case, each object has multiple target categories.

Definition 7.10 (Bounded or Unlimited Number of Tasks) In the bounded case, a budget is given and only a number of tasks can be asked. In this case, it aims to find the category as accurately as possible within the budget. In the unlimited case, it aims to ask the minimal number of questions to precisely identify the target categories.

Definition 7.11 (Structured Category) The categories are not independent and can be organized as a tree (or forest, which is a set of trees) or a directed acyclic graph (DAG).

Parameswaran et al. [67] give the complexities of the category problems for all combinations. They further prove that the problems with the DAG model are NP-hard and the problems with the tree model can be solved in polynomial time. For the DAG model, they design brute-force algorithms to find the optimal solutions; and for the tree model, they propose dynamic programming algorithms.

A more complicated categorization problem is building a taxonomy using crowdsourcing, which is mainly studied in the HCI community. A pioneer work is the Cascade system [18]. The input of the Cascade system consists of a set of items to be categorized and a descriptive phrase describing a topic of the items. An example is 100 response to the question "What is your best travel advice" with the topic "Travel Advice." The output is a taxonomy with labeled categories and associated items. In the previous example, a taxonomy of "Travel Advice" is a tree consisting categories such as "air travel" and "saving money," and a response of travel tips about flight may be associated with these categories. Note that in Cascade, both the categories and the association between items and categories are obtained from crowdsourcing. Cascade introduces a crowd algorithm that produces a global understanding of the items and updates the taxonomy as new data arrives. To support this, it proposes HIT primitives and global structure inference method that combines independently generated judgments to these HITs.

7.7 Crowdsourced Skyline

Existing works often adopt the definition of *Crowdsourced Skyline Query* as follows.

Definition 7.12 (Crowdsourced Skyline Query) Given a collection of items, a skyline operator aims to find all the items in the collection that are not *dominated* by others.

Based on the above definition, it is often assumed that each item has n attributes, and we say an item o_1 is dominated by another item o_2 if and only if o_1 is not as good as o_2 for any attribute. For example, suppose we want to find the best hotel in a city by considering three attributes: price, distance, and rating. There might be a lot of hotels in the city. Using a skyline operator, we can remove those hotels that cannot be our optimal choice.

Machine-based skyline algorithms have been extensively studied for more than a decade [12]. However, there are still some limitations that are not easy to overcome without human involvement. For example, when data has missing values, it is difficult to obtain high-quality skyline results without asking humans to fill those missing values; when data contains some attributes that are hard to compare (e.g., deciding which photo looks more beautiful), we need to ask humans to make such comparisons. These limitations motivate two research directions for crowdsourced skyline.

7.7.1 Crowdsourced Skyline on Incomplete Data

One direction is to study how to leverage crowdsourcing to obtain high-quality skyline answers from incomplete data. It is expensive to ask the crowd to fill all missing values. In fact, some missing values are not necessary to be filled. For example, suppose we want to know whether an item A dominates another item B. If we can derive from incomplete data that B has one attribute that is better than A, then we will know that A does not dominate B, without filling any missing value. Existing works [55, 56] adopt a hybrid human-machine workflow, which first uses machines to fill all missing values in a heuristic way, and then asks the crowd to examine those items that are most beneficial to improve the skyline quality.

The basic idea for crowdsourced skyline computation on incomplete data is missing data prediction. In order to minimize the error brought by prediction, the authors in [55] introduced a concept of *prediction risk*, which implies the risk that each tuple may have on the final result of skyline query once the predicted value is applied without the use of crowdsourcing. Based on the computation of such a risk factor, the crowdsourcing efforts can be restricted to the tuples that will strongly affect the quality of skyline result while just using reasonable predicted values for the remaining tuples since they may only have a limited impact to the quality of skyline result.

In order to reach this goal, [55] designs a framework. The main goal is to design a system that is self-tuning, i.e., for each database instance, the best parameters are self-tuned heuristically while preserving quality of the skyline result. The reason of such design is due to the fact that the quality of prediction methods often varies a lot w.r.t. the distributions of the data. Thus, it introduces a toolbox that combines different implementations of strategies for various components. In general, for the tuples with missing values, the predictions will be made before computation. Then, for high-risk tuples, the predicted values will be updated based on the crowdsourcing results.

7.7.2 Crowdsourced Skyline with Comparisons

Another idea that incorporates crowdsourcing into the skyline computation is to use the crowd to compare those challenging attributes. Because the crowd workers are very likely to make mistakes, the skyline algorithm has to handle noisy comparison answers. Groz and Milo [35] studied how many comparisons are required to return the correct skyline with a high probability.

Next, we illustrate an example of skyline comparison, which tries to find the best cities by considering two attributes: salaries and education qualities. Note that the missing values are inherent in databases and thus the salaries or education qualities may be missing. Also, it may be hard to retrieve numerical estimated values from the crowd. However, compared to the numerical estimated values, the comparisons of rankings should be more natural and feasible to ask the crowd. Thus, in order to compute the skyline query, the crowd may help to answer the comparison tasks such as "Is the education better in city x than city y?" or "Is the salary better in city x compared to city y?." The crowd may make mistakes; thus, each task will be assigned to multiple workers, and the workers' answers are aggregated to obtain the final result.

In the work [35], the authors assume that the crowd workers will make errors with some probability, and thus the quality of the aggregated result will be increased by assigning a task to more workers. The goal is to minimize the number of comparisons or assignments of each task and at the same time make sure that the correct result of the skyline query will be returned with a high probability. To this end, they design methods that make use of the potentially small size of the skyline and analyze the number of comparisons required to obtain a decent quality [35]. The authors also make contributions to predicting the most likely skyline query given the partial information of noisy comparisons, and they also show that the optimal prediction is indeed computationally intractable.

7.8 Crowdsourced Planning

With the power of crowd workers, existing works [47, 78, 82, 106, 109] focus on addressing the *crowdsourced planning query*, which is defined as follows.

Definition 7.13 (Crowdsourced Planning Query) Crowdsourced planning generally tries to design a sequence of actions from an initial state to a final state or to reach a goal.

For example, starting at the position of Harvard University, a traveler seeks ways to UC Berkeley. In another example, a student wants to plan the sequence of courses to take in a semester, in order to obtain a solid knowledge of a specific domain. We first introduce a general crowdsourced planning query [47] and then focus on specific planning applications that were studied in [106] and [78], respectively.

7.8.1 General Crowdsourced Planning Query

The crowdsourced planning query is defined in [47] as a query whose output is a sequence of objects or actions that gets one from some initial state to some ideal target state. For example, a vocation planning query aims at finding a sequence of places to visit in a vocation, with the goal of enjoying the vocation the most. Different from traditional planning query which is automatically solved by computers in the artificial intelligence field [62], crowdsourced planning query is more difficult in the involved computational complexity (e.g., the ordering of all possible visiting places) and the hard-to-understand goals (e.g., enjoy vocation the most) by computers, while they are relatively easier to be understood by humans.

The general problem defined in [47] is shown as follows.

Definition 7.14 (Crowdsourced Planning Query) Given a large set of known items, the target is to choose a subset of items and then order the chosen items in a sequence that gives the best value (the "value" is hard to formalize, but they can be comprehended by humans).

To be specific, suppose a vocation query is to design a plan that *will visit several cities in Italy, starting from the city of Rome*. Then the set of known items S can be constructed as the input, say $S = \{Rome,\ Florence,\ Milan,\ Naples,\ Venice\}$. An ordered list of items form a plan p, say $p = (Rome,\ Naples,\ Milan)$. Note that the set of all possible plans (or ordered lists) that can be formed from the set S can be modeled as a tree T. Each node in T is an item (in S), and it has only one ancestor, i.e., the item preceding it in the plan, while it has multiple children, i.e., the items that follow it in different plans. A tree can be incrementally built by asking workers to answer some tasks. The tasks used in [47] are used to seek for the subsequent items in a (partial) plan. That is, given a possible plan p, a task would like to ask the subsequent item that can be added in p. For example, "Given the sequence $(Rome,\ Florence,\ Naples)$, what is the next city that I should visit?". If the answer "Milan" is given, then an edge $(Naples,\ Milan)$ will be added to T (if not existed in T).

In order to control the budget (while guaranteeing quality), [47] restricts each task to be asked at most N times (N is predetermined), and then based on the answers, each edge (u, v) in T is assigned with a score indicating the probability that the answer is v, by considering all tasks that ask about the subsequent item in the plan ending at u. For example, if $N = 10$ and in the answers asking for the subsequent item of the sequence $(Rome,\ Florence,\ Naples)$, suppose eight workers answer with $Milan$ and two workers answer with $Venice$, then the edge $(Naples,\ Milan)$ has a score of 0.8, while the edge $(Naples,\ Venice)$ has a score of 0.2. The target of the problem is to obtain a "good" (or optimal) plan. To capture the goodness of a plan p, the scores of edges in p are used. As each score in an edge indicates the degree of consent of the subsequent item to be added in the sequence, the score of a plan p is defined as the product of the scores of all edges in p. When the tree is completed (i.e., no extra edge can be added), a final tree T_f is built, and the optimal path p_f can be calculated as the path with highest score in T_f.

To trade-off cost and quality, [47] formally defines that a plan p is a *correct answer* to the query if $score(p_f) - score(p) \leq \epsilon$. That is, as a tree is incrementally built, the target is to find a satisfying plan p with its score at least $score(p_f) - \epsilon$. As a final tree T_f is not completely built, finding a correct answer (i.e., a satisfying plan) requires to enumerate all possible completions of trees given the current tree T. One completion of T should consider the expansion of T by answering possible tasks as well as the possible answers given by the tasks. By leveraging the idea, [47] designs a greedy algorithm to decide which task should be the next to ask and then decides the stopping conditions to ask tasks. Before introducing the ways in [47] on how to solve the above two problems, some notations are clarified: a score for a path p in a tree T is denoted as $score_T(p)$; given a tree T, all the possible generated final trees form a set $Final(T)$; given a possible final tree T', all the paths starting from the root node to leaf nodes form a set $Set(T')$.

(1) Which Task to Ask Next: To decide the next task to ask, given current tree T, it selects the path that is the most possible to derive a path with a high score. In order to reach this goal, there are three steps:

- Step 1: For each path p in the tree T, it computes the potential score for p, indicating the highest score that p can reach in all the possible final trees:

$$\max{}_{T' \in Final(T)}\ score_T(p);$$

- Step 2: It selects a plan p in the tree T that has the highest potential score;
- Step 3: Based on the selected plan, the task asks the path where the starting node is the root node, and the ending node is the closest-to-the-root not-yet-exhausted node.

The task of asking tasks in batch is also considered in [47], which corresponds to the scenario that multiple workers may come at the same time. Heuristics are proposed, for example, the case that each task is asked N times is considered to assign multiple same tasks in parallel to the coming individual workers.

(2) Stopping Conditions: To check whether it is okay to stop asking tasks, [47] defines a notion of uncertainty, representing the maximal difference between a score of given sequence and the highest sequence score in all possible final trees. That is, in a tree T, the uncertainty U of a path p starting at root node and ending at leaf node is defined as

$$U(T, p) = \max_{T' \in Final(T)} \left[\max_{p' \in Set(T')} score_{T'}(p') - score_{T'}(p) \right].$$

And the stopping criteria is to decide whether or not the uncertainty is lower than a predefined threshold, i.e., $U(T, p) < \epsilon$.

Note that the cost, which is related to the number of tasks being asked, is also considered in [47]. Inspired by [28], the notion of *instance-optimal* is used, i.e., for any possible input, the cost is of the same magnitude compared with any other correct algorithm. Finally, [47] proves that the developed algorithm is instance-optimal and the optimality ratio is far by at most a factor of two from the lower bound.

7.8.2 An Application: Route Planning

Route planning is a problem mainly studied in [78, 106], defined as follows: starting at a point, what is the best route to the target point? The best route should consider the real-time conditions, e.g., weather conditions, road conditions, etc. Recently, there are two works [78, 106] that study the problem in parallel. Next, we discuss them, respectively.

7.8.2.1 Crowdsourced Route Selection [106]

Given starting and target points, existing systems can help generate multiple routes. As observed in [106], too many generated routes may destroy the usability of recommended data and at the same time decrease the user satisfaction. Based on the observation, [106] proposes ways to reduce the uncertainties of generated routes. As the target is to select the best route, [106] regards the input of the problem as a set of generated routes, where each route has a known probability to be the best route.

Let us give an example to illustrate such process. The starting point is denoted as s, and target point is denoted as t, and there are four routes generated with probabilities where each route i is denoted as R_i, with its probability to be best route denoted as $\Pr(R_i)$. For example, suppose there are four routes generated:

Route R_1 : $s - v_1 - t$, with $\Pr(R_1) = 0.1$;
Route R_2 : $s - v_2 - v_4 - t$, with $\Pr(R_2) = 0.2$;
Route R_3 : $s - v_3 - v_4 - t$, with $\Pr(R_3) = 0.4$;
Route R_4 : $s - v_3 - t$, with $\Pr(R_4) = 0.3$.

To capture the uncertainties of all routes, [106] uses *Shannon entropy*, as it has some rationale consistent with the hardness of selecting the best route. To be specific, given a probability distribution $(p_1, p_2, \ldots, p_n)$, the entropy is defined as $-\sum_{i=1}^{n} p_i \cdot \log p_i$. Intuitively, the lower the entropy, the more skewed is the distribution (or the probability concentrates in one index); the higher the entropy, the distribution tends to be of uniform distribution. The entropy of the distribution defined on the best route in the example is calculated as

$$-(0.1 \cdot \log 0.1 + 0.2 \cdot \log 0.2 + 0.4 \cdot \log 0.4 + 0.3 \cdot \log 0.3) = 1.28.$$

As we want to select the best route, which indicates that the values of probability should concentrate in one route, thus the tasks should be selected such that the workers' answers can help to decrease the entropy.

To construct tasks, by considering the simplicity of tasks, [106] uses routing query (RQ), which asks workers the directions of an intersection. To be precise, the worker is given (1) a point v (it can be s), (2) the target point t, and (3) a few choices (a few points that conform a set D) adjacent to v, and the task is that *starting at v, which point in D should I go in order to arrive at t*. For the same example, suppose s ="HKUST," t ="HKU," and v_1, v_2, and v_3 are "Hang Hau," "Choi Hung," and "Kowloon Bay," respectively, a RQ task is "Which direction should I go from HKUST (s) to HKU (t), v_1, v_2, or v_3?".

Then the problem in [106] is formally defined as follows.

Definition 7.15 (Crowdsourced Routing Query) Given a best route set and a budget B indicating the number of RQs, within the budget design strategies that crowdsource RQs in order to maximally reduce the entropy defined on the best route probability distribution.

Note that as a worker gives answers, the best route distribution is changed, resulting the change in entropy. Suppose workers always give true answers, and with the above RQ task, the worker answers with point v_3, then it means that the next step of s is v_3, so routes R_1 and R_2 contradicting the right direction should be discarded. Then the best route distribution only leaves routes R_3 and R_4. To form a probability distribution, their probabilities are normalized as $\Pr(R_3) = 0.4/(0.4 + 0.3) = 0.57$, $\Pr(R_4) = 0.3/(0.4 + 0.3) = 0.43$. Then the entropy is changed to

$$-(0.57 \cdot \log 0.57 + 0.43 \cdot \log 0.43) = 0.683,$$

resulting in reducing the entropy. As the worker's answer is known, there are multiple RQs that can be crowdsourced. In order to estimate the benefit of each RQ, [106] uses the expected entropy reduction to measure the goodness of a RQ in reducing uncertainty. That is, it assumes that a worker will have a probability to give a specific answer to a RQ, which can be derived from $\Pr(R_1)$ to $\Pr(R_4)$. Finally, [106] will select the RQ which has the highest expected entropy reduction as the task to assign.

Besides, [106] also considers the case that worker may have some qualities to correctly answer a task. Then the utilization of crowdsourced answers with answering qualities is taken into consideration. Moreover, [106] makes two extensions: (1) *Assign multiple RQs to a worker*. Although selecting the optimal k RQs is NP-hard, the submodularity property is observed, and efficient greedy method is proposed to give $(1 - 1/e)$-approximation ratio guarantee. (2) *Consider anther type of task, i.e., binary routing query (BRQ)*. BRQ is similar to RQ, as it only wants to know whether one direction is okay or not, so only two answers "Yes" and "No" are provided. For example, a BRQ task is "From HKUST (s) to HKU (t), should I go via v_1 or not?". This type of task considers the case that a road may have multiple intersections and a worker may only be familiar with some of them.

7.8.2.2 Crowdsourced Route Recommendation [78]

Previous researches show that even the route recommendation provided by the widely used service providers can deviate from experienced drivers' choices significantly, because of the fact that many latent factors may affect drivers' choices, such as traffic lights, speed limits, road conditions, etc. Similar to [106], here [78] also takes some candidate routes as input and leverage workers (experienced drivers) to help select the best route. To address the problem of selecting the best route, [78] specifically deals with the following two important problems:

Task generation: how to automatically generate user-friendly tasks?
Worker selection: how to choose a set of workers for a given task?

For *task generation*, [78] adopts the idea of utilizing significant locations (i.e., landmarks) in order to help describe a route in high level, which can be easily understood by humans. By using landmarks, the differences in various routes can be explicitly and vividly presented to a worker, and the worker feels more comfortable to answer the tasks. For the task type, [78] uses the simple task encoded in landmarks with possible answers "Yes" and "No." An example task is: "Do you prefer the route passing landmark A at 2:00 pm?" The landmark related to a task can discriminate some routes from other tasks. The task generation takes three steps:

Step 1:   *Inferring landmark significance.* It infers the significance of each landmark by considering worker's similarity with it. The inference in [78] leverages online check-in records from location-based social network and trajectories of vehicles.

Step 2:   *Landmark selection.* [78] takes three criteria to select landmarks. First, the landmarks should have high significance (such that workers are familiar with them); second, the landmarks should be discriminative among the routes; and third, the size of selected landmarks should be as small as possible (in order to reduce the workload of workers). By adopting these criteria, the problem is to select a small set of high significant landmarks that are discriminative to all routes.

Step 3:   *Task ordering.* Having selected the landmarks, [78] considers the following two intuitions in ordering tasks to a worker: (1) the next task to ask should be based on the answer of the previous task, and thus the task order takes a hierarchical structure; (2) the landmark should be significant and the information gain of asking the task should be high.

For *worker selection*, [78] selects a set of eligible workers by considering the following aspects: (1) *Response time.* Given a task, a worker should give the answer before the specified response time. The response time for a worker can be modeled as an exponential distribution. If the learned distribution shows that the worker has a low probability to give an answer before the specified response time, then the task will not be assigned to the worker. (2) *Worker's familiarity score.* To get a better quality for a task, the worker's familiarity with the task's landmark is important. For each worker, [78] will evaluate the familiarity with each landmark using two factors:

the worker's profile information (e.g., home address) and the history of worker's finished tasks around the landmark's region.

Then to find the top-k eligible workers, [78] selects k workers with the best knowledge of the task. The knowledge is defined on both the familiarity score of the landmarks in a task and the coverage of the knowledge.

7.9 Crowdsourced Schema Matching

The well-known *schema matching problem* is commonly defined as finding the attribute mappings between different database schemas. For example, given two relational tables: Table *Professor* with attributes *(Name, Phone)* and Table *ProfInfo* with attributes *(Profname, Fax, Tel)*, the target is to find a correct matching between the attributes in the two tables. A possible matching is {*(Professor.Name, ProfInfo.Profname), (Professor.Phone, ProfInfo.Tel)*}. Existing works [72] develop automatic algorithms to generate a set of possible matchings, by considering linguistic and structural information. However, it is rather hard to remove the ambiguity using machine-only approaches. Thus, recent works [29, 64, 105] leverage human insights (or crowdsourcing) to reduce the uncertainty in schema matching.

Hybrid Human-Machine Method Given two schemas, Zhang et al. [105] propose a hybrid (human-machine) approach called CrowdMatcher to address the schema matching problem. Specifically, they first use machine-based approaches to generate a set of possible matchings, where each matching (containing a set of correspondences) has a probability to be the correct matching. Then they define the uncertainty in schema matching as the entropy of all possible matchings. To reduce the uncertainty, they utilize the crowdsourcing. Since these workers are not expert, the tasks answered by them cannot be hard. Therefore, each task is to ask whether or not a given correspondence is a correct matching. For example, "Is the correspondence *(Professor.Name, ProfInfo.Profname)* correct?". In order to balance cost and quality, they develop effective task selection techniques, with the target of reducing the most uncertainty of schema matching with the lowest cost. Figure 7.11 illustrates the system architecture of CrowdMatcher.

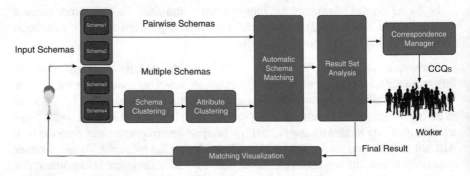

Fig. 7.11 Schema matching system architecture

Firstly, two or more schemas are taken as an input of the system. If there are only two schemas, the input of the automatic schema matching module is the two schemas. Otherwise, they start with the schema clustering module and then generate the mediated schema for every cluster in the attribute clustering module. Then they use the automatic schema matching module to perform the top-k schema matching. Based on the matching results, they generate the correspondence set with the result set analysis module. Next the set will pass to correspondence manager, which selects optimal correspondence correctness queries (CCQs), which are ready to be answered by the crowd to determine whether a given correspondence is correct or not. After workers answer the questions, the answers are returned back to the result set analysis module to adjust the probability distribution, and the module then selects another batch of CCQs. Finally, the results are returned to the requester through a matching visualization module when the time or budget is exhausted.

Pay-as-you-go Method Hung et al. [64] consider a pay-as-you-go setting where the matching should satisfy the specified integrity constraints. And the matching can also be also conducted in a network of multiple schemas. The built network conforms to some specified integrity constraints. For example, the 1-to-1 constraint, which forces that each attribute of one schema can be matched to at most one attribute of another schema; the cycle constraint, if matching schemas are in a cycle, the matching schemas must form a cycle too. As Fig. 7.12 shows, $\{c_3, c_5\}$ violates the 1-to-1 constraint and $\{c_1, c_2, c_5\}$ violates the cycle constraint. Figure 7.13 illustrates the framework of this work. Firstly, the schema matchers generate some correspondence candidates automatically. Then in order to reduce the uncertainty of the whole schema network, they first build a probabilistic matching network, where each attribute correspondence is associated with a probability. The uncertainty of the schema network is defined as the entropy of a set of random variables, where each one denotes whether a given correspondence exists in the correct matching. They take the probabilistic network as an input of the instantiation module, which generates an approximate selective matching result. In order to improve the quality of instantiation, they propose an algorithm to select correspondences to ask human experts, so as to reduce the uncertainty of the schema network the most.

Different from relational tables which give relatively complete schema information, Fan et al. [29] study how to match web tables, which are rather dirty and

Fig. 7.12 1-to-1 and cycle constraints

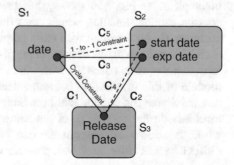

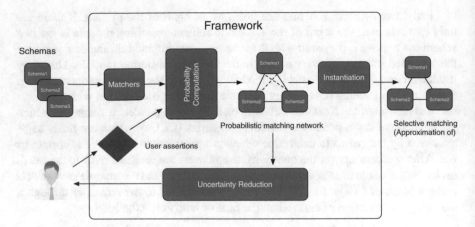

Fig. 7.13 Pay-as-you-go schema matching

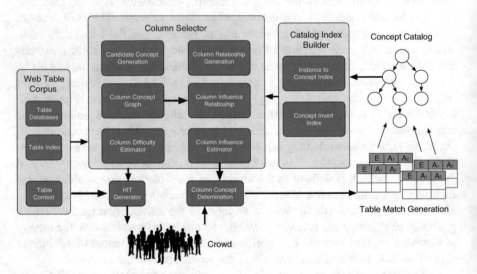

Fig. 7.14 Hybrid machine-crowdsourcing web table matching

incomplete. As Fig. 7.14 shows, they leverage the power of knowledge bases (the concept catalog module), which can offer concepts with wide coverage in a high accuracy. Given the input web table corpus, their approach first maps the values in each table column to one or more concepts, and the columns (in different tables) that represent the same concept are matched with each other. Then the most important module of this framework, column selector, selects columns to ask crowd, which can maximize the benefit within a budget. Then, they adopt a hybrid approach: machines do the easy work of matching pairs of columns based on the concepts, while the crowd will discern the concepts for the columns that are regarded as difficult by machines. Each task presents workers with values in a column as well as a set of candidate concepts, and then workers will decide which is the best concept for the column. In terms of task selection, a utility function is defined over columns, considering the factors such as the columns' difficulties, and effective solutions are proposed next, targeting at selecting which columns to ask so that the expected utility is maximized.

7.10   Crowd Mining

Crowd mining tries to learn and observe significant patterns based on workers' answers. The foundations of crowd mining is firstly developed in [8], and then the following works [4–6, 10] enrich the area of crowd mining from different perspectives. We first introduce the crowd mining formulation [8] in Sect. 7.10.1 and then discuss the complexity of crowd mining [4] in Sect. 7.10.2. Finally, we, respectively, introduce two approaches that leverage the power of ontologies (or knowledge bases) to enhance crowd mining [6] and [10] in Sect. 7.10.3.

7.10.1   Crowd Mining Formulation

Based on the fact that journalists, markets, and politicians routinely find new trends in analyzing people's behaviors and put forward new policies, the problem of discovering significant patterns in crowd's behavior is an important but challenging task. To capture the significant patterns, association rule [3] in data mining area is used, which tries to find when one set of items indicates the presence of another set of items. For example, the analysis of customers' buying behaviors shows that the purchases of diapers are closely associated with the purchases of beer. The reason is that when fathers arrive at shopping center to buy diapers for their babies, they would buy beer as well since they could not go to pubs as often (as they have babies). The reason why traditional approach cannot be directly applied to crowdsourcing setting is that the association rule mining depends on the fact that it mines from a database with lots of transactions (e.g., list of buying items for each customer); however, in crowdsourcing area, one cannot reply a worker to provide extensive such transactions (e.g., to tell all their past activities).

Another example is that a health researcher is interested in analyzing the performance of traditional medicine and he tries to discover the association rules such that "Garlic can be used to treat flu." In this case, he can neither count on a database which only contains symptoms and treatments for a particular disease nor ask the healers for an exhaustive list of all the cases that have been treated. But social studies have shown that *although people cannot recall all their transactions, they can provide simple summaries (or called "personal rules") to the question.* For example, they may know that "When I have flu, most of the cases I will take Garlic because it indeed useful to me." Given personal rules answered by different persons, they can be aggregated together to find an overall important rule (or the general trends). So the problem in [8] is defined as follows.

Definition 7.16 (Crowd Mining Rule Finding) Based on the personal rules collected from crowd workers, seek for a way to aggregate them and find the overall important rules (i.e., general trends).

First, an association rule (denoted as r) is defined as $r : A \rightarrow B$ indicating that A implies B. An example rule r : flu→garlic means that *if having flu, then take garlic*. There are two statistics defined on a rule $r : A \rightarrow B$: **support** (r) and **confidence** (r). The support of rule r indicates the probability that A and B occur together, and confidence of rule r indicates that given A occurs, the conditional probability B occurs. Given a rule r, its support and confidence (simple summaries) can be answered by workers. For example, given a r : flu→garlic, a task asking about the support of r goes like this "How often do you usually have flu and use garlic as the treatment?"; a task asking about the confidence of r goes like this "When you have the flu, how often do you take garlic?." Then, workers will give their personal support s and confidence c for the rule r. A worker's answer may be "Once a Week," "Twice a Day." etc., and these answers can be encoded by probabilities. For example, two workers' answers for a rule r can be $(0.4, 0.6)$ and $(0.6, 0.4)$, where each worker's answer is indicated by a tuple $(support, confidence)$.

In order to address the problem, given workers' answers, there are still two important questions:

(1) **Aggregation problem**: how to compute the significant rules based on workers' answers?
(2) **Assignment problem**: which rule should be chosen as the next task to assign when a worker comes?

For the aggregation problem, [8] proposes estimation-based methods to estimate the support and confidence. We next illustrate the solutions to some of the important steps.

- *Estimating the mean*: When a task is asked about the support and confidence for a rule r, the process can be modeled as choosing a random worker from a huge number of workers and obtain his support and confidence. Based on the central limit theorem, the sample mean, denoted as $f_r(s, c)$, follows a normal distribution, with means that μ and variance $\frac{1}{N}\Sigma$, where N is the number of answers collected from workers, and μ, Σ are, respectively, mean and covariance of the answers given by workers. We have

$$f_r \sim \mathcal{N}\left(\mu, \frac{1}{N}\Sigma\right),$$

which will be used in the following steps.
- *Estimating rule significance*: In order to capture whether a rule is significant or not, remember a rule is defined based on two probabilities: the support s and confidence c. The larger the two probabilities are, the more significant the rule

is. So two other threshold parameters θ_s and θ_c are defined. If the support s and confidence c are both greater than the respective defined thresholds, i.e., $s \geq \theta_s$ and $c \geq \theta_c$, the rule can be regarded as a significant rule. As the sample mean conforms to a normal distribution, the problem becomes to compute the integral of f_r as $s \geq \theta_s$ and $c \geq \theta_c$:

$$sig(r) = \int_{\theta_s}^{\infty} \int_{\theta_c}^{\infty} f_r(s, c) \, dc \, ds,$$

which indicates the probability that rule r is significant based on current worker's answers. So to identify whether a rule r is significant or not, one can set a threshold ϵ (say 0.5) and see whether $sig(r) \geq \epsilon$. Thus, the aggregation problem has been addressed.

For the assignment problem, the idea in [8] is to select the most beneficial rule to ask. The benefit of a rule r is calculated as the expected quality of improvement if the rule r is asked. Initially, [8] defines the current quality of rule r. Then in order to capture a worker's answer, the distribution of a worker's answer of the rule r is computed. Based on considering the worker's answer of each rule, the next quality is computed, and all the rules are ranked based on the quality of improvement. Finally, the rules with highest ranks can be assigned. Next, we illustrate the solutions to some of the important steps as follows.

- *Estimating current quality*: For each rule r, we identify it as significant or not based on the probability $sig(r)$ defined above. So the current quality of a rule r is defined as the error rate $P_{err}(r)$ or probability that the rule is wrongly identified. To be specific, if a rule r is regarded as significant, then the error rate is the probability that the rule r is actually not significant, i.e., $1 - sig(r)$; otherwise, if a rule r is regarded as not significant, then the error rate is the probability that the rule r is actually significant, i.e., $sig(r)$. So the current error rate $P_{err}(r)$ is

$$P_{err}(r) = sig(r) \cdot \mathbb{1}_{\{r \text{ is not identified as significant}\}}$$
$$+ [1 - sig(r)] \cdot \mathbb{1}_{\{r \text{ is identified as significant}\}},$$

and the current quality can be regarded as $1 - P_{err}(r)$.
- *Estimating sample distribution*: As defined above, the sample mean f_r conforms to a normal distribution. But here we are interested in the distribution of the next sample, defined as g_r. Based on current answers for a rule r, it can be briefly modeled as $g_r \sim \mathcal{N}(\mu, \Sigma)$. But note that the sampled data (s, c) should satisfy probability constraint, $s \in [0, 1]$ and $c \in [0, 1]$, and then g_r is actually modeled as the truncated Gaussian, i.e.,

$$g_r \sim \mathcal{N}(\mu, \Sigma) \text{ and } g_r((s, c) \mid s, c \in [0, 1]).$$

- *Estimating next quality*: Knowing the sample distribution above, [8] uses a Monte Carlo method. It is a repeated process, where in each process:

(1) a sample (s, c) is drawn from g_r as defined above;
(2) based on considering the new sample (s, c), a new value $f'_r \sim \mathcal{N}(\mu', \frac{1}{N+1}\Sigma')$ is computed, where μ' and Σ' are updated mean and variance by considering the new sample;
(3) based on the new f'_r, the significance $sig(r)$ is updated as

$$sig'(r) = \int_{\theta_s}^{\infty} \int_{\theta_c}^{\infty} f'_r(s, c) \, dc \, ds,$$

and the error rate P_{err} is update as

$$P'_{err}(r) = sig'(r) \cdot \mathbb{1}_{\{r \text{ is identified as significant}\}}$$
$$+ [1 - sig'(r)] \cdot \mathbb{1}_{\{r \text{ is not identified as significant}\}}.$$

We keep $P'_{err}(r)$ and by repeating the processes, the average error, denoted as $avg(P'_{err}(r))$, can be computed, indicating the expected error rate of rule r if the next task of r is answered. The value $1 - avg(P'_{err}(r))$ can be regarded as the next quality.

- *Final ranking of rules*: For a rule r, based on the current quality $1 - P_{err}(r)$, and the next quality $1 - avg(P'_{err}(r))$, the quality of improvement can be represented as

$$1 - avg(P'_{err}(r)) - (1 - P_{err}(r)) = P_{err}(r) - avg(P'_{err}(r)),$$

which is indeed the error rate reduction. So the higher the error rate is reduced, the higher the quality is improved, and the higher rank the rule r is. Thus, rules with high ranks will be assigned, which addresses the assignment problem.

Having addressed the above two important problems, [8] also considers the problem that making computations on the entire domains of rules is very expensive (which is also a challenging task in traditional associate rule finding problem). As workers are mostly unable to remember rules with large sizes, so [8] proposes heuristics to consider rules with small sizes, without considering their dependencies. Moreover, other than only using the task asking about the support and confidence of a rule, [8] also proposes another type of task: open task, which asks the worker to provide a possible rule with statistics. For example, "Tell about an illness, the way you treat it, and how often the both (illness and treatment) occur." [8] considers the trade-off between asking open tasks to obtain new possible information or to ask tasks of existing rules to improve the estimation.

7.10.2 Complexity of Crowd Mining

Given the formulation of crowd mining specified in [8], the theoretical foundations of crowd mining is studied in [4]. Similar to the problem in [8] which finds significant rules (using support and confidence), in [4], the complexity of a specific problem, i.e., mining frequent itemset, is studied. The frequent itemset is defined on a subset of items and study whether they are frequent enough in crowd's behaviors. An example frequent itemset is $A = \{sport, tennis\}$, meaning that sport and tennis usually occur together, i.e., people often play tennis as the sport. The definition of frequent itemset uses a similar definition as the support of a rule. To make it simple, given a rule $r : A \rightarrow B$, the frequency of an itemset $A \cup B$ is just defined as the support of rule r or the probability that all items in the itemset $A \cup B$ occur. An itemset A is considered as frequent if its support s exceeds a predefined threshold $\theta_s \in [0, 1]$, i.e., $s > \theta_s$. Thus, a frequency function $freq(\cdot)$ is defined, which takes an itemset A as input, outputs *true* if A is frequent, and outputs *false* if A is not frequent. The problem is then formally defined as follows.

Definition 7.17 (Frequent Itemset Mining) Given a set of items $\mathcal{I}$, select all subsets of items such that for each subset $A \subseteq \mathcal{I}$, it should satisfy $freq(A) = true$.

To measure the complexity, there are three problems: (1) how to define the relationship between items; (2) how to model the crowd; (3) what complexities should be considered. Next, we discuss these three problems, respectively.

(1) *The relationship between items*: [4] uses taxonomy, and the reason is that a taxonomy can represent the semantic relations between items, which conform to the properties of items. To be specific, a taxonomy is a partial "is-a" relationship on the items and is defined as a partially ordered set $\Psi = (\mathcal{I}, \leq)$, where $\leq$ is a partial order representing the relation "is-a." For example, "sport$\leq$tennis" means that tennis is a sport (or tennis is more specific than sport). An example taxonomy is $\Psi = \{$sport$\leq$tennis, sport$\leq$football, tennis$\leq$racket$\}$.

(2) *Model the crowd*: To make it simple, [4] models the crowd query which takes an input as an itemset $A \subseteq \mathcal{I}$, and the output is $freq(A)$, i.e., *true* or *false*. The simple model is set for analysis, which regards that workers always give true answers. An example query is "Do you usually play tennis as a sport?" and the answer is *Yes* or *No* indicating the true answer.

(3) *Complexities to be studied*: Given the taxonomy Ψ and the worker model, [4] studies two complexities: **crowd complexity** and **computational complexity**. The first measures the number of questions that need to be asked in order to compute the frequent itemsets, while the second measures the computational effort required to choose the questions. There is a trade-off between the these two complexities: investing more computational complexity to select questions may reduce the budgets to ask the crowd, while giving more budgets in asking questions may reduce the complexity to select questions. [4] studies the complexities of the two problems under the defined model theoretically.

7.10.3 Leveraging Ontology in Crowd Mining

To leverage the ontology (knowledge bases) in crowd mining, we first describe the
work [6] which builds a system OASSIS (for Ontology Assisted Crowd Mining)
and then [10] talk about how to integrate the natural language processing to
OASSIS. Finally, a generic architecture of crowd mining [5], considering the general
knowledge of ontology and the individual knowledge of the crowd, is explained.

7.10.3.1 OASSIS: Ontology Assisted Crowd Mining [6]

Consider the following scenario where a mother would like to bring her child and go
for a vocational tour in NYC, and then the objective is to *find popular combinations
of an activity in a child-friendly attraction in NYC and a restaurant nearby.*

   Although web-search can be used to answer such query, the words child-friendly
attraction and nearby restaurant are hard to be encoded. And it is time-consuming
to filter the returned text-based documents. Even though the query can be posted to
forums, the text-based replies also take time to aggregate. Based on this intuition, [6]
proposes to develop a system that enables requesters to pose general queries to the
crowd, such that the relevant answers representing frequent patterns can be returned.

   In order to build such system, [6] combines the ontology, which can provide
general knowledge, and the crowd, which can offer personal knowledge together.
Next, we discuss the **ontology** and the **crowd**, respectively.

Ontology The ontology (or knowledge base) contains a number of facts (c_1, r, c_2)
in a graph form, where c_1 and c_2 are objects (nodes in the graph) and r is a relation
(edges in the graph). An example fact is (Central_park, inside, NYC) indicating
that Central Park is inside NYC. Another example fact gives hierarchical meaning
(Biking, SubClassof, Activity), which indicates that biking is a type of (or belongs
to the category of) activity. Enriched with ontology, there are many "universal truth"
related to the query that can be used.

Crowd The crowd's behavior can be encoded by facts. For example, a fact (Biking,
doAt, Central_Park) means that a person has biked in the Central Park. Multiple
facts form a fact-set indicating the actions that a particular person has done in
a certain period. For example, a fact-set {(Biking, doAt, Central_Park), (Falafel,
eatAt, Maoz_Veg)} means that a person has biked at Central Park and eaten Falafel
at Maoz_Veg in a certain period of time. Then a person is encoded by a personal
database containing a set of fact-sets indicating all his past behaviors.

   But the problem is that the database is "virtual," i.e., we cannot count on the
person to recall all his past behaviors, and what we can do is to ask a person to
provide basic summaries (or the support) for some fact-set (just as what [8] does).
So the problem becomes to *find frequent fact-sets that can meet the requester's
query*, where the frequency of a fact-set can be defined as the average support for all
workers. Following this idea, [6] designs two types of task that can ask the crowd:

(1) **Concrete task** Which tries to find the support of a fact-set for a worker. For example, for the fact-set {(Biking, doAt, Central_Park), (Falafel, eatAt, Maoz_Veg)}, the task is "How often do you go biking in Central Park and eat Falafel at Maoz_Veg?."

(2) **Specification task** Given a fact-set, it tries to ask the crowd to give more facts enriching the fact-set and at the same time provide its support. For example, for the fact-set {(Biking, doAt, Central_Park), (Falafel, eatAt, Maoz_Veg)}, the task is "Other than going biking in Central Park and eating Falafel at Maoz_Vegwhat, what type of sport do you also do in Central Park? And how often do you do that?."

In order to reduce the number of tasks that need to be asked, the ontology can be used. For example, if doing sports in Central Park is regarded as infrequent, then it is not worth asking the crowd the support of going biking in Central Park. Moreover, the ontology can also help to derive a redundant set of answers. For example, if going biking is frequently done in Central Park, then more general facts (e.g., doing sports) are also frequent.

Finally, [6] develops a query language called OASSIS-QL based on the SPARQL. It provides an interface for requesters and is a requester-friendly query formulation tool. As it is adapted from SPARQK, well-developed tools translating natural language to SPARQL can be used. The query language to the expression (or natural language query) "find popular combinations of an activity in a child-friendly attraction in NYC and a restaurant nearby" can be represented as follows:

```
SELECT fact-sets
WHERE
{w subClassof Attraction.
x instanceOf w.
x inside NYC.
x hasLabel "child-friendly".
y subClassOf Activity.
z instanceOf Restaurant.
z nearBy x}
SATISFYING
{y doAt x.   [ ]² eatAt z}
```

with support $= 0.4$. A possible answer for this query is $x =$ Central_Park, $w =$ Park, $y =$ Biking, and $z =$ Maoz_Veg, which indicates the answer that you can go biking in Central Park and eat at Maoz_Veg restaurant. The SELECT statement tries to select the answers (frequent fact-sets), or some targeted variables and the WHERE statement indicate the general knowledge in the natural language query, defining a SPARQL-like query on topology. The SATISFYING statement indicates individual knowledge in the natural language query, defining the patterns to be mined from the crowd (corresponding to the asked tasks) and patterns with

---

²the symbol "[]" means *anything*, and its value is not cared.

the support (computed based on different workers' answers) satisfying a preset
threshold will be returned.

7.10.3.2   The Natural Language Processing Component in OASSIS [10]

Remember that in OASSIS [6], it develops the query language of OASSIS-QL and
would like the requester to build such kind of formal and declarative query (as the
above example shows) and feed to the system of QASSIS. But this may be too
hard for most requesters without technical backgrounds. Different from [6], in [10],
it considers the problem of how to translate the natural language (NL) query to a
formal, declarative OASSIS-QL query, by considering both the general knowledge
and individual knowledge behind the NL.

Take an example NL query as follows: "What are the most interesting places near
Forest Hotel, Buffalo, we should visit in the fall?" Note that to answer such query,
both the general and individual knowledge are required: (1) The general knowledge
contains places in Buffalo and the proximity to Forest Hotel, which can be answered
from the ontology, (2) while the individual knowledge contains the degree of interest
to a place and the places that should be visited in the fall, which can be obtained from
workers. So the objective defined in [10] is to develop a principal approach to the
translation of NL queries that can combine general and individual knowledge, into
formal OASSIS-QL queries.

The general approach adopted by [10] to address the problem can be summarized
as the following five steps.

Step 1.   Given a NL query, the NL processing tools (e.g., the dependency
parser [60]) are used to create a dependency graph. The dependency graph
is a tree, where each node is a term (e.g., "What," "are," etc.) and each edge
represents the semantic dependency between terms (e.g., the edge visit→should
be labeled by "auxiliary").

Step 2.   Based on the generated dependency graph, the individual expressions
(IX)-detector, developed by [10], tries to classify the dependency graph into
two parts: one part corresponds to the individual knowledge and the other
part corresponds to the general knowledge. The basic idea of IX-detector is to
perform a pattern-based detection by considering two criteria: (1) using training
sentences and machine learning technologies to learn; (2) using predefined
structural patterns and vocabularies. The patterns are designed based on three
typical sources of individuality: (1) lexical individuality (a term that has a lexical
meaning), e.g., the term "interesting" in "interesting places"; (2) participant
individuality (the event that includes an participant), e.g., the term "we" in "we
should visit in the fall"; (3) syntactic individuality (a syntactic structure), e.g.,
the term "should" in "we should visit in the fall". Finally two parts generated
by IX-detector (i.e., general knowledge part and individual knowledge part) are
dealt with, which will be introduced below.

Step 3.   For the general knowledge part, the existing general query generators
can help translate the NL query (corresponding to the general knowledge) to
SPARQL query. For example, suppose the NL of the general knowledge part
is "What places are near Forest Hotel, Buffalo?" After the process of existing
general query generators tools, the SPARQL query is generated as

```
SELECT DISTINCT x
WHERE
{x instanceOf Place.
x near Forest_Hotel,_Buffalo,_NY}
```
Finally, the two triples $(x, \text{instanceOf}, \text{Place})$ and $(x\ \text{near}\ \text{Forest\_Hotel,\_Buffalo,}$
$\_\text{NY})$ are generated.

Step 4.   For the individual knowledge part, the IX-translator developed by [10]
defines some rules about how to translate each possible pattern in the dependency
tree to a OASSIS-QL triple. For example, the pattern $(x, \text{prep\_in}, y)$ corresponds
to the triple $([], \text{in}, y)$, and then the edge visit$\rightarrow$fall (with label "prep_in")
in the dependency graph will be translated to $([], \text{in}, \text{fall})$. Taking the above
example, there are two parts of the dependency graph that conform to the
individual language, i.e., *the most interesting* and *we should visit in the fall*. They
correspond to the triples $(x, \text{hasLabel interesting})$, $([], \text{visit}, x)$, and $([], \text{in}, \text{fall})$.

Step 5.   Based on the triples generated by the above two steps, the query compo-
sition developed by [10] will combine them into a formal OASSIS-QL query.
The OASSIS-QL query defined in [6] contains three statements: SELECT,
WHERE, and SATISFYING. The WHERE statement corresponds to the general
knowledge part, and the SATISFYING statement corresponds to the individual
knowledge part. For the SATISFYING statement, [10] divides it into several
subclauses, where each subclause corresponds to a single event (e.g., *we should
visit in the fall*) or property (*the most interesting*). The words "most," "best," etc.
will be translated to a top-k (k is set based on different intensities) selection in
the corresponding subclause; otherwise, a predefined support threshold is used.

Based on the above five steps, the formal OASSIS-QL query based on the NL
query in the example is generated:

```
SELECT VARIABLES x
WHERE
{x instanceOf Place.
x near Forest_Hotel,_Buffalo,_NY}
SATISFYING
{x hasLabel "interesting"}
ORDER BY DESC(SUPPORT)
LIMIT 5
AND
{[ ] visit x.
[ ] in Fall}
WITH SUPPORT THRESHOLD = 0.1.
```

7.10.3.3   A Generic Architecture of Crowd Mining [5]

A generic architecture for crowd mining is described in [5], which can help the researchers compare between existing crowdsourcing systems and point out future extensions. Next, we summarize different components of the generic architecture as follows:

Data Repositories   The three main data repositories are knowledge base, the user profile, and crowd results:

(1) Knowledge base is the main data repository storing long-term data, which contains two parts: the input data and inferred data. The input data integrates existing ontologies (knowledge bases), conventional databases, etc. The inferred data is based on the analysis of crowd input. For example, for the general data, the crowd can assist in filling in the missing values (e.g., the missing address of a company), while for the individual data, the crowd can offer their personal opinions or basic summaries.

(2) the user profile contains the detailed information of each worker, and there are two parts: the input data and inferred data. The input data contains the registered information to the system (e.g., the user's name and date of birth). The inferred data is derived from a worker's past performance. For example, based on a worker's answering history, we observe the worker's interest, or his quality, etc.

(3) the crowd results is a temporary repository which contains the crowd's answers for a particular query. It both contains the crowd's raw answers and additional inference results based on crowd's answers.

Query Processing Engine   Given a declarative query, the objective of the query engine is to compute an efficient plan and conduct query execution. It can be summarized as to minimize the crowd effort and the machine execution time. Note that some parts of a query can be derived from the existing data (e.g., the knowledge base abovementioned), but other parts rely on the crowd's input.

Crowd Task Manager   Given the crowd's answers for a task, the target of crowd task manager is to aggregate the results. For the general knowledge task, the aggregation can identify the correct answer with a high accuracy, while for the individual knowledge task, things are different. As the answers for the task tend to be open (each worker has their own perspectives), the aggregation tends to use a significant function to estimate whether certain result is significant or not based on sufficient workers' answers. The crowd task manager also handles the assignment issue. It selects tasks based on considering two factors: (1) the estimated improvement of utility if the task is assigned and answered and (2) the recent answering history of active workers (it tends to select relative tasks based on his answered tasks).

Inference and Summarization   This part deals with raw answers collected from the crowd. It makes use of the existing data (e.g., knowledge base) to do inference on the current worker's answers and may restore the inferred data into the data repositories. Then the subsequent executions can benefit from the executions of previous tasks.

Crowd Selection The component will select appropriate workers to assign tasks. The selection will consider three factors: (1) the explicit preferences given by the requester, (2) the similarity between the requester and workers, and (3) worker's properties for the task, e.g., the quality and expertise in the related area. Moreover, another responsibility for the component is to filter spammer or malicious workers.

NL Parser/Generator Given the natural language (NL) query of a worker, the part will automatically generate a declarative query for the query engine. It should handle both the general and individual knowledge information in the NL well. Moreover, once the tasks are assigned to workers and workers' answers are collected, these two processes may also use to translate from natural language (NL) to machine-readable language.

7.11 Spatial Crowdsourcing

Many crowdsourced tasks contain spatial information, e.g., checking whether there is a parking slot and taking a photo of a restaurant to check whether there are available seats. Spatial crowdsourcing has a significant difference from other operators: the workers can answer most of the tasks in other operators, while workers can only answer some spatial tasks whose locations are close to the workers. Thus, most of existing studies focus on how to effectively assign the tasks to appropriate workers based on spatial proximity [11, 17, 24, 38, 41–43, 48, 49, 71, 79, 83, 106]. Another important application is taxi-hailing (e.g., Uber), where there are drivers and riders and we need to assign drivers to appropriate drivers (based on the distance).

Based on the scenarios on how they define spatial distance, we discuss them in two dimensions: (1) Euclidean space and (2) road network.

7.11.1 Spatial Crowdsourcing with Euclidean Space

In Euclidean space, the locations of both spatial tasks and workers are represented by geo-coordinates with longitudes and latitudes. Spatial tasks can be assigned to workers with two different modes [48]: (1) worker selection model, where the spatial crowdsourcing server (SC-server) publicly publishes the tasks on a crowdsourcing platform and online workers can voluntarily choose the nearby spatial tasks; (2) server assignment model, where all online workers send their locations to the SC-server and the SC-server assigns the tasks to workers.

(1) Worker Selection Model

The advantage of the worker selection task mode is easy to implement. Deng et al. [24] study the *maximum task scheduling* problem, which aims to recommend a longest valid sequence of tasks for a worker. Given a worker and a set of tasks where each task has an expiration time, a worker is asked to do a sequence of tasks. To be specific, after finishing one task, the worker is required to travel to the location of the next task, which may incur some time cost. The goal is to find the longest valid sequence for the worker that all the tasks in the sequence can be completed within the expiration time. The problem is proved to be NP-hard, and a dynamic programming algorithm is proposed to find the exact optimal answer with time complexity of $\mathcal{O}(2^n \cdot n^2)$, where n is the number of available tasks. Since the exact algorithms cannot scale to large number of tasks, effective approximation algorithms have been proposed by greedily picking the tasks with the least expiration time or the smallest distance.

However, this model has some limitations. First, the SC-server does not have any control over tasks and workers, which may result in the case that some spatial tasks will never be assigned, while others are assigned redundantly. Second, a single worker does not have a global knowledge on all the tasks, and thus tasks are blindly selected without considering the overall travel cost and task expiration time.

(2) Server Assignment Model

The server assignment model overcomes the limitations of worker selection model, which assigns tasks to nearby workers while optimizing some overall objectives.

Maximize the Number of Assigned Tasks [48] Kazemi et al. [48] assume that every worker has a spatial region and can accept at most T tasks within the region. For each time instance t, it runs an assignment algorithm to assign available tasks to workers under above constraints. Given a time interval consisted of several continuous time instances $t_1, t_2, \cdots, t_n$, it aims to maximize the total number of assigned tasks during the time interval. The problem is proved to be NP-hard. To solve the problem, it first proposes a greedy strategy to do the maximum assignment at every time instance by reducing the problem to the maximum flow problem on a transformed graph. Then it improves the performance by giving high priorities to the tasks with high location entropy during the assignment, where a high location entropy indicates that there are many workers around the area of the task which can be more easily assigned.

Maximize the Number of Correctly Assigned Tasks [49] Kazemi et al. [49] assume that (a) each task has a confidence score and (b) each worker has a reputation score and a group of workers have an aggregated reputation score, which can be computed based on the reputation scores of individual workers in the group. The problem aims to maximize the number of tasks assigned to workers or groups, satisfying that (a) the task is in the spatial region of the worker; (b) for a task, the reputation score of the assigned worker or the aggregation score of the assigned group must be larger than the confidence score; and (c) a worker can only accept at most T tasks. The problem is proved to be NP-hard, and a greedy algorithm is proposed by iteratively assigning tasks to workers or groups until no more valid assignments exist. It also proposes an incremental search strategy which removes a single task-worker (task-group) assignment generated by the greedy algorithm and replaces it with assignments of more tasks. The incremental search process is computationally expensive, and some heuristics are proposed to improve its efficiency.

Maximize Successful Rate [38] Note that some tasks may not be able to be successfully completed. For example, a worker may give up a task, and the task can only be completed successfully with a certain probability. Hassan et al. [38] aim to maximize the successful rate under limited budget. The problem is analogous to the well-known multi-armed bandit problem where each worker is considered as an arm and each assignment is equivalent to paying an arm. The probability of being successful is the same as the resulting reward. It extends the traditional multi-armed bandit algorithms to the spatial assignment problem.

A drawback of the server assignment model is that workers should report their locations to the SC-server, which can pose privacy issues [71, 83].

7.11.2 Spatial Crowdsourcing with Road Network

There are several works on road network. We have shown that [79] and [106] study the route planning problem in Sect. 7.8. Artikis et al. [11] utilize spatial crowdsourcing to manage urban traffic on road network, where the crowdsourcing component is used to supplement the data sources by querying human volunteers. If two sensors disagree on a traffic situation, e.g., the same bus within a small time period reports two different congestion degrees, they will search workers close to the location with disagreements and ask them about the real traffic situation. Ta et al. [81] studied the spatial task assignment problem under road-network constraints and proposed optimization techniques to reduce the complicated operations of computing road-network distance.

7.12 Summary of Crowdsourced Operators

This section summarizes the existing studies on crowdsourced operators in Table 7.1, including the adopted techniques, used task types, and optimization goals. As shown in the table, one needs to design different techniques to optimize the operators.

Table 7.1 Crowdsourced operators

Operators		Task type	Goal	Techniques
Selection	Filtering [65, 66]	Single choice	Quality	Answer aggregation, task assignment
			Cost	Task selection
	Find [76]	Single choice	Quality	Answer aggregation, task assignment
			Cost	Task selection
			Latency	Multi-batch latency control
	Search [100]	Single choice	Quality	Answer aggregation, task assignment
			Cost	Task selection
			Latency	Single-batch latency control
Collection	Enumeration [84]	Labeling	Quality	Answer aggregation
			Cost	Miscellaneous (Pay-as-you-go approach)
	Fill [68]	Labeling	Quality	Answer aggregation
			Cost	Miscellaneous (Compensation scheme)
			Latency	Single-batch latency control
Join	CrowdER [89]	Single choice & Clustering	Quality	Worker elimination, answer aggregation
			Cost	Pruning, miscellaneous (Task design)
	Transitivity [88, 94, 95]	Single choice	Quality	Answer aggregation, task assignment
			Cost	Pruning, answer deduction
	[33, 98]	Single choice	Quality	Answer aggregation, task assignment
			Cost	Task selection
Topk/Sort	Heuristics-based [37]	Single choice	Quality	Answer aggregation, task assignment
			Cost	Task selection
	Machine learning [16, 69]	Single choice	Quality	Answer aggregation, task assignment
			Cost	Task selection
	Iterative reduce [27, 37]	Single choice	Quality	Answer aggregation, task assignment
			Cost	Task selection, answer deduction
	Heap-based [20]	Single choice	Quality	Answer aggregation, task assignment
			Cost	Task selection, answer deduction
	Hybrid [50, 101]	Single choice & rating	Quality	Answer aggregation, task assignment
				Task selection

Categorize	[67]	Single choice	Quality	Answer aggregation, task assignment
			Cost	Task selection
Aggregation	Max [37, 85]	Single choice	Quality	Answer aggregation, task assignment
			Cost	Task selection, answer deduction
	Count [58]	Single choice, Labeling	Quality	Worker elimination, answer aggregation
			Cost	Sampling, miscellaneous (Task design)
	Median [39]	Single choice	Quality	Answer aggregation
			Cost	Sampling
	Group by [20]	Single choice	Quality	Answer aggregation, task assignment
			Cost	Task selection, answer deduction
Skyline	[55, 56]	Labeling	Quality	Answer aggregation, task assignment
			Cost	Task selection
	[35]	Single choice	Quality	Answer aggregation, task assignment
			Cost	Task selection
Planning	CrowdPlanr [47, 57]	Labeling	Quality	Answer aggregation, task assignment
			Cost	Task selection, answer deduction
	Route planning [106]	Single choice	Quality	Answer aggregation, task assignment
			Cost	Task selection, answer deduction
	CrowdPlanner [78, 79]	Single choice	Quality	Answer aggregation, task assignment
			Cost	Task selection
Schema Matching	[29, 64, 105]	Single choice	Quality	Answer aggregation, task assignment
			Cost	Task selection, answer deduction
Mining	CrowdMiner [8, 9]	Labeling	Quality	Answer aggregation, task assignment
			Cost	Task selection
	OASSIS [6, 7]	Labeling	Quality	Answer aggregation, task assignment
			Cost	Task selection, answer deduction
Spatial	[83]	Labeling	Quality	Answer aggregation, task assignment
			Cost	Task selection

References

1. Amazon mechanical turk. https://www.mturk.com/
2. Adelsman, R.M., Whinston, A.B.: Sophisticated voting with information for two voting functions. Journal of Economic Theory **15**(1), 145–159 (1977)
3. Agrawal, R., Imielinski, T., Swami, A.N.: Mining association rules between sets of items in large databases. In: SIGMOD, pp. 207–216 (1993)
4. Amarilli, A., Amsterdamer, Y., Milo, T.: On the complexity of mining itemsets from the crowd using taxonomies. In: ICDT, pp. 15–25 (2015)
5. Amsterdamer, Y., Davidson, S., Kukliansky, A., Milo, T., Novgorodov, S., Somech, A.: Managing general and individual knowledge in crowd mining applications. In: CIDR (2015)
6. Amsterdamer, Y., Davidson, S.B., Milo, T., Novgorodov, S., Somech, A.: Oassis: query driven crowd mining. In: SIGMOD, pp. 589–600. ACM (2014)
7. Amsterdamer, Y., Davidson, S.B., Milo, T., Novgorodov, S., Somech, A.: Ontology assisted crowd mining. PVLDB **7**(13), 1597–1600 (2014)
8. Amsterdamer, Y., Grossman, Y., Milo, T., Senellart, P.: Crowd mining. In: SIGMOD, pp. 241–252. ACM (2013)
9. Amsterdamer, Y., Grossman, Y., Milo, T., Senellart, P.: Crowdminer: Mining association rules from the crowd. PVLDB **6**(12), 1250–1253 (2013)
10. Amsterdamer, Y., Kukliansky, A., Milo, T.: Nl2cm: A natural language interface to crowd mining. In: SIGMOD, pp. 1433–1438. ACM (2015)
11. Artikis, A., Weidlich, M., Schnitzler, F., Boutsis, I., Liebig, T., Piatkowski, N., Bockermann, C., Morik, K., Kalogeraki, V., Marecek, J., et al.: Heterogeneous stream processing and crowdsourcing for urban traffic management. In: EDBT, pp. 712–723 (2014)
12. Börzsönyi, S., Kossmann, D., Stocker, K.: The skyline operator. In: ICDE, pp. 421–430 (2001)
13. Bradley, R.A., Terry, M.E.: Rank analysis of incomplete block designs: I. the method of paired comparisons. Biometrika pp. 324–345 (1952)
14. Busa-Fekete, R., Szorenyi, B., Cheng, W., Weng, P., Hullermeier, E.: Top-k selection based on adaptive sampling of noisy preferences. In: ICML, pp. 1094–1102 (2013)
15. Chai, C., Li, G., Li, J., Deng, D., Feng, J.: Cost-effective crowdsourced entity resolution: A partial-order approach. In: SIGMOD, pp. 969–984 (2016)
16. Chen, X., Bennett, P.N., Collins-Thompson, K., Horvitz, E.: Pairwise ranking aggregation in a crowdsourced setting. In: WSDM, pp. 193–202 (2013)
17. Chen, Z., Fu, R., Zhao, Z., Liu, Z., Xia, L., Chen, L., Cheng, P., Cao, C.C., Tong, Y., Zhang, C.J.: gmission: a general spatial crowdsourcing platform. PVLDB **7**(13), 1629–1632 (2014)
18. Chilton, L.B., Little, G., Edge, D., Weld, D.S., Landay, J.A.: Cascade: crowdsourcing taxonomy creation. In: CHI, pp. 1999–2008 (2013). doi:10.1145/2470654.2466265
19. Chung, Y., Mortensen, M.L., Binnig, C., Kraska, T.: Estimating the impact of unknown unknowns on aggregate query results. In: SIGMOD, pp. 861–876 (2016). doi:10.1145/2882903.2882909
20. Davidson, S.B., Khanna, S., Milo, T., Roy, S.: Using the crowd for top-k and group-by queries. In: ICDT, pp. 225–236 (2013)
21. Demartini, G., Difallah, D.E., Cudré-Mauroux, P.: Zencrowd: leveraging probabilistic reasoning and crowdsourcing techniques for large-scale entity linking. In: WWW, pp. 469–478 (2012)
22. Deng, D., Li, G., Hao, S., Wang, J., Feng, J.: Massjoin: A mapreduce-based method for scalable string similarity joins. In: ICDE, pp. 340–351 (2014)
23. Deng, D., Li, G., Wen, H., Feng, J.: An efficient partition based method for exact set similarity joins. PVLDB **9**(4), 360–371 (2015)
24. Deng, D., Shahabi, C., Demiryurek, U.: Maximizing the number of worker's self-selected tasks in spatial crowdsourcing. In: SIGSPATIAL, pp. 324–333. ACM (2013)

25. Deng, D., Tao, Y., Li, G.: Overlap set similarity joins with theoretical guarantees. In: SIGMOD (2018)
26. Elo, A.E.: The rating of chessplayers, past and present, vol. 3. Batsford London (1978)
27. Eriksson, B.: Learning to top-k search using pairwise comparisons. In: AISTATS, pp. 265–273 (2013)
28. Fagin, R., Lotem, A., Naor, M.: Optimal aggregation algorithms for middleware. Journal of Computer and System Sciences **66**(4), 614–656 (2003)
29. Fan, J., Lu, M., Ooi, B.C., Tan, W.C., Zhang, M.: A hybrid machine-crowdsourcing system for matching web tables. In: ICDE, pp. 976–987. IEEE (2014)
30. Fan, J., Wei, Z., Zhang, D., Yang, J., Du, X.: Distribution-aware crowdsourced entity collection. IEEE Trans. Knowl. Data Eng. (2017)
31. Feige, U., Raghavan, P., Peleg, D., Upfal, E.: Computing with noisy information. SIAM J. Comput. pp. 1001–1018 (1994)
32. Feng, J., Wang, J., Li, G.: Trie-join: a trie-based method for efficient string similarity joins. VLDB J. **21**(4), 437–461 (2012)
33. Gokhale, C., Das, S., Doan, A., Naughton, J.F., Rampalli, N., Shavlik, J.W., Zhu, X.: Corleone: hands-off crowdsourcing for entity matching. In: SIGMOD, pp. 601–612 (2014)
34. Gomes, R., Welinder, P., Krause, A., Perona, P.: Crowdclustering. In: NIPS, pp. 558–566 (2011)
35. Groz, B., Milo, T.: Skyline queries with noisy comparisons. In: PODS, pp. 185–198 (2015)
36. Gruenheid, A., Kossmann, D., Ramesh, S., Widmer, F.: Crowdsourcing entity resolution: When is A=B? Technical report, ETH Zürich
37. Guo, S., Parameswaran, A.G., Garcia-Molina, H.: So who won?: dynamic max discovery with the crowd. In: SIGMOD, pp. 385–396 (2012)
38. ul Hassan, U., Curry, E.: A multi-armed bandit approach to online spatial task assignment. In: UIC (2014)
39. Heikinheimo, H., Ukkonen, A.: The crowd-median algorithm. In: HCOMP (2013)
40. Herbrich, R., Minka, T., Graepel, T.: Trueskill: A bayesian skill rating system. In: NIPS, pp. 569–576 (2006)
41. Hu, H., Li, G., Bao, Z., Feng, J.: Crowdsourcing-based real-time urban traffic speed estimation: From speed to trend. In: ICDE, pp. 883–894 (2016)
42. Hu, H., Li, G., Bao, Z., Feng, J., Wu, Y., Gong, Z., Xu, Y.: Top-k spatio-textual similarity join. IEEE Trans. Knowl. Data Eng. **28**(2), 551–565 (2016)
43. Hu, H., Zheng, Y., Bao, Z., Li, G., Feng, J.: Crowdsourced poi labelling: Location-aware result inference and task assignment. In: ICDE, pp. 61–72 (2016)
44. Jeffery, S.R., Franklin, M.J., Halevy, A.Y.: Pay-as-you-go user feedback for dataspace systems. In: SIGMOD, pp. 847–860 (2008)
45. Jiang, X., Lim, L.H., Yao, Y., Ye, Y.: Statistical ranking and combinatorial hodge theory. Math. Program. pp. 203–244 (2011)
46. Jiang, Y., Li, G., Feng, J., Li, W.: String similarity joins: An experimental evaluation. PVLDB **7**(8), 625–636 (2014)
47. Kaplan, H., Lotosh, I., Milo, T., Novgorodov, S.: Answering planning queries with the crowd. PVLDB **6**(9), 697–708 (2013)
48. Kazemi, L., Shahabi, C.: Geocrowd: enabling query answering with spatial crowdsourcing. In: SIGSPATIAL, pp. 189–198. ACM (2012)
49. Kazemi, L., Shahabi, C., Chen, L.: Geotrucrowd: trustworthy query answering with spatial crowdsourcing. In: SIGSPATIAL, pp. 304–313 (2013)
50. Khan, A.R., Garcia-Molina, H.: Hybrid strategies for finding the max with the crowd. Tech. rep. (2014)
51. Li, G., Deng, D., Feng, J.: A partition-based method for string similarity joins with edit-distance constraints. ACM Trans. Database Syst. **38**(2), 9:1–9:33 (2013)
52. Li, G., Deng, D., Wang, J., Feng, J.: PASS-JOIN: A partition-based method for similarity joins. PVLDB **5**(3), 253–264 (2011)

53. Li, G., He, J., Deng, D., Li, J.: Efficient similarity join and search on multi-attribute data. In: SIGMOD, pp. 1137–1151 (2015)
54. Li, K., Li, X.Z.G., Feng, J.: A rating-ranking based framework for crowdsourced top-k computation. In: SIGMOD, pp. 1–16 (2018)
55. Lofi, C., Maarry, K.E., Balke, W.: Skyline queries in crowd-enabled databases. In: EDBT, pp. 465–476 (2013)
56. Lofi, C., Maarry, K.E., Balke, W.: Skyline queries over incomplete data - error models for focused crowd-sourcing. In: ER, pp. 298–312 (2013)
57. Lotosh, I., Milo, T., Novgorodov, S.: Crowdplanr: Planning made easy with crowd. In: ICDE, pp. 1344–1347. IEEE (2013)
58. Marcus, A., Karger, D.R., Madden, S., Miller, R., Oh, S.: Counting with the crowd. PVLDB 6(2), 109–120 (2012)
59. Marcus, A., Wu, E., Karger, D.R., Madden, S., Miller, R.C.: Human-powered sorts and joins. PVLDB 5(1), 13–24 (2011)
60. de Marneffe, M.C., MacCartney, B., Manning, C.D.: Generating typed dependency parses from phrase structure parses. In: LREC (2006)
61. Mozafari, B., Sarkar, P., Franklin, M., Jordan, M., Madden, S.: Scaling up crowd-sourcing to very large datasets: a case for active learning. PVLDB 8(2), 125–136 (2014)
62. Nau, D., Ghallab, M., Traverso, P.: Automated Planning: Theory & Practice. Morgan Kaufmann Publishers Inc., San Francisco, CA, USA (2004)
63. Negahban, S., Oh, S., Shah, D.: Iterative ranking from pair-wise comparisons. In: NIPS, pp. 2483–2491 (2012)
64. Nguyen, Q.V.H., Nguyen, T.T., Miklós, Z., Aberer, K., Gal, A., Weidlich, M.: Pay-as-you-go reconciliation in schema matching networks. In: ICDE, pp. 220–231. IEEE (2014)
65. Parameswaran, A.G., Boyd, S., Garcia-Molina, H., Gupta, A., Polyzotis, N., Widom, J.: Optimal crowd-powered rating and filtering algorithms. PVLDB 7(9), 685–696 (2014)
66. Parameswaran, A.G., Garcia-Molina, H., Park, H., Polyzotis, N., Ramesh, A., Widom, J.: Crowdscreen: algorithms for filtering data with humans. In: SIGMOD, pp. 361–372 (2012)
67. Parameswaran, A.G., Sarma, A.D., Garcia-Molina, H., Polyzotis, N., Widom, J.: Human-assisted graph search: it's okay to ask questions. PVLDB 4(5), 267–278 (2011)
68. Park, H., Widom, J.: Crowdfill: collecting structured data from the crowd. In: SIGMOD, pp. 577–588 (2014)
69. Pfeiffer, T., Gao, X.A., Chen, Y., Mao, A., Rand, D.G.: Adaptive polling for information aggregation. In: AAAI (2012)
70. Pomerol, J.C., Barba-Romero, S.: Multicriterion decision in management: principles and practice, vol. 25. Springer (2000)
71. Pournajaf, L., Xiong, L., Sunderam, V., Goryczka, S.: Spatial task assignment for crowd sensing with cloaked locations. In: MDM, vol. 1, pp. 73–82. IEEE (2014)
72. Rahm, E., Bernstein, P.A.: A survey of approaches to automatic schema matching. VLDBJ 10(4), 334–350 (2001)
73. Rekatsinas, T., Deshpande, A., Parameswaran, A.G.: Crowdgather: Entity extraction over structured domains. CoRR abs/1502.06823 (2015)
74. Sarawagi, S., Bhamidipaty, A.: Interactive deduplication using active learning. In: SIGKDD, pp. 269–278 (2002)
75. Sarma, A.D., Jain, A., Nandi, A., Parameswaran, A., Widom, J.: Jellybean: Crowd-powered image counting algorithms. Technical report, Stanford University
76. Sarma, A.D., Parameswaran, A.G., Garcia-Molina, H., Halevy, A.Y.: Crowd-powered find algorithms. In: ICDE, pp. 964–975 (2014)
77. Shang, Z., Liu, Y., Li, G., Feng, J.: K-join: Knowledge-aware similarity join. IEEE Trans. Knowl. Data Eng. 28(12), 3293–3308 (2016)
78. Su, H., Zheng, K., Huang, J., Jeung, H., Chen, L., Zhou, X.: Crowdplanner: A crowd-based route recommendation system. In: ICDE, pp. 1144–1155. IEEE (2014)
79. Su, H., Zheng, K., Huang, J., Liu, T., Wang, H., Zhou, X.: A crowd-based route recommendation system-crowdplanner. In: ICDE, pp. 1178–1181 (2014)

80. Sun, J., Shang, Z., Li, G., Deng, D., Bao, Z.: Dima: A distributed in-memory similarity-based query processing system. PVLDB **10**(12), 1925–1928 (2017)
81. Ta, N., Li, G., Feng, J.: An efficient ride-sharing framework for maximizing shared route. TKDE **32**(9), 3001–3015 (2017)
82. Talamadupula, K., Kambhampati, S., Hu, Y., Nguyen, T.A., Zhuo, H.H.: Herding the crowd: Automated planning for crowdsourced planning. In: HCOMP (2013)
83. To, H., Ghinita, G., Shahabi, C.: A framework for protecting worker location privacy in spatial crowdsourcing. PVLDB **7**(10), 919–930 (2014)
84. Trushkowsky, B., Kraska, T., Franklin, M.J., Sarkar, P.: Crowdsourced enumeration queries. In: ICDE, pp. 673–684 (2013)
85. Venetis, P., Garcia-Molina, H., Huang, K., Polyzotis, N.: Max algorithms in crowdsourcing environments. In: WWW, pp. 989–998 (2012)
86. Verroios, V., Garcia-Molina, H.: Entity resolution with crowd errors. In: ICDE, pp. 219–230 (2015)
87. Verroios, V., Garcia-Molina, H., Papakonstantinou, Y.: Waldo: An adaptive human interface for crowd entity resolution. In: SIGMOD, pp. 1133–1148. ACM (2017)
88. Vesdapunt, N., Bellare, K., Dalvi, N.N.: Crowdsourcing algorithms for entity resolution. PVLDB **7**(12), 1071–1082 (2014)
89. Wang, J., Kraska, T., Franklin, M.J., Feng, J.: CrowdER: crowdsourcing entity resolution. PVLDB **5**(11), 1483–1494 (2012)
90. Wang, J., Li, G., Feng, J.: Trie-join: Efficient trie-based string similarity joins with edit-distance constraints. PVLDB **3**(1), 1219–1230 (2010)
91. Wang, J., Li, G., Feng, J.: Fast-join: An efficient method for fuzzy token matching based string similarity join. In: ICDE, pp. 458–469 (2011)
92. Wang, J., Li, G., Feng, J.: Can we beat the prefix filtering?: an adaptive framework for similarity join and search. In: SIGMOD, pp. 85–96 (2012)
93. Wang, J., Li, G., Feng, J.: Extending string similarity join to tolerant fuzzy token matching. ACM Trans. Database Syst. **39**(1), 7:1–7:45 (2014)
94. Wang, J., Li, G., Kraska, T., Franklin, M.J., Feng, J.: Leveraging transitive relations for crowdsourced joins. In: SIGMOD, pp. 229–240 (2013)
95. Wang, S., Xiao, X., Lee, C.: Crowd-based deduplication: An adaptive approach. In: SIGMOD, pp. 1263–1277 (2015)
96. Wauthier, F.L., Jordan, M.I., Jojic, N.: Efficient ranking from pairwise comparisons. In: ICML, pp. 109–117 (2013)
97. Weng, X., Li, G., Hu, H., Feng, J.: Crowdsourced selection on multi-attribute data. In: CIKM, pp. 307–316 (2017)
98. Whang, S.E., Lofgren, P., Garcia-Molina, H.: Question selection for crowd entity resolution. PVLDB **6**(6), 349–360 (2013)
99. Whang, S.E., McAuley, J., Garcia-Molina, H.: Compare me maybe: Crowd entity resolution interfaces. Technical report, Stanford University
100. Yan, T., Kumar, V., Ganesan, D.: Crowdsearch: exploiting crowds for accurate real-time image search on mobile phones. In: MobiSys, pp. 77–90 (2010)
101. Ye, P., EDU, U., Doermann, D.: Combining preference and absolute judgements in a crowd-sourced setting. In: ICML Workshop (2013)
102. Yi, J., Jin, R., Jain, A.K., Jain, S., Yang, T.: Semi-crowdsourced clustering: Generalizing crowd labeling by robust distance metric learning. In: NIPS, pp. 1781–1789 (2012)
103. Yu, M., Li, G., Deng, D., Feng, J.: String similarity search and join: a survey. Frontiers of Computer Science **10**(3), 399–417 (2016)
104. Yu, M., Wang, J., Li, G., Zhang, Y., Deng, D., Feng, J.: A unified framework for string similarity search with edit-distance constraint. VLDB J. **26**(2), 249–274 (2017)
105. Zhang, C.J., Chen, L., Jagadish, H.V., Cao, C.C.: Reducing uncertainty of schema matching via crowdsourcing. PVLDB **6**(9), 757–768 (2013)
106. Zhang, C.J., Tong, Y., Chen, L.: Where to: Crowd-aided path selection. PVLDB **7**(14), 2005–2016 (2014)

107. Zhang, X., Li, G., Feng, J.: Crowdsourced top-k algorithms: An experimental evaluation. PVLDB **9**(4) (2015)
108. Zhuang, Y., Li, G., Zhong, Z., Feng, J.: Hike: A hybrid human-machine method for entity alignment in large-scale knowledge bases. In: CIKM, pp. 1917–1926 (2017)
109. Zhuo, H.H.: Crowdsourced action-model acquisition for planning. In: AAAI, pp. 3439–3446

Chapter 8
Conclusion

Crowdsourcing has become more and more prevalent in data management and analytics. This book provides a comprehensive review of crowdsourced data management, including motivation, applications, techniques, and existing systems. This chapter first summarizes this book in Sect. 8.2 and then provides several research challenges and opportunities in Sect. 8.2.

8.1 Summary

In this book, we extensively review the recent studies on crowdsourced data management. We first summarize the crowdsourcing overview and workflow and focus on micro-tasks and the worker payment model. Most existing algorithms emphasize balancing quality, cost, and latency, and we summarize existing techniques to address these challenges. For quality control, we review worker modeling, worker elimination, truth inference, and task assignment techniques; for cost control, we discuss machine-based pruning technique, task selection, answer deduction, sampling, and task-design techniques; for latency control, we discuss single-task, single-batch, and multi-batch latency control and introduce different latency control models. Next, we review existing crowdsourced data management systems and optimization techniques. We also discuss crowdsourced operators and summarize the techniques to support various operators.

8.2 Research Challenges

In this section, we discuss some research challenges and opportunities in crowdsourced data management.

© Springer Nature Singapore Pte Ltd. 2018
G. Li et al., *Crowdsourced Data Management*,
https://doi.org/10.1007/978-981-10-7847-7_8

8.2.1 Query Plan

Because SQL is a declarative query language, a single query often corresponds to multiple query plans; and thus it relies on a query optimizer to select the best plan. Traditionally, the way a query optimizer works is to estimate the computation cost of each query plan and choose the one with the minimum estimated cost. However, this process turns to be quite challenging in a crowdsourcing environment because (1) there are three optimization objectives (result quality, monetary cost, and latency) that need to be considered together and (2) humans are much more unpredictable than machines.

8.2.2 Benchmark

A large variety of TPC benchmarks (e.g., TPC-H for analytic workloads, TPC-DI for data integration) standardize performance comparisons for database systems and promote the development of database research. Although there are some open datasets (http://dbgroup.cs.tsinghua.edu.cn/ligl/crowddata), there is still lack of standardized benchmarks available. In order to better explore the research topic, it is important to study how to develop evaluation methodologies and benchmarks for crowdsourced data management systems.

8.2.3 Big Data

In the big data era, data volumes are increasing very fast. Compared to machines, humans are much more expensive, and thus it would be increasingly more costly to apply crowdsourcing to emerge big data scenarios. There are some existing works that aim to address this problem, but they only work for some certain data processing tasks, such as data cleaning [6] and data labeling [5]. Therefore, it is important to continue this study and to develop new techniques that work for all kinds of data processing tasks.

8.2.4 Macro-tasks

Most of existing studies focus on micro-tasks, which can be easily assigned to workers and instantly answered by workers. However many real applications need to use macro-tasks, such as writing a paper. Macro-tasks are hard to be split and accomplished by multiple workers, because they will lose the context information if they are split [4]. Workers are not interested in answering a whole macro-task as each macro-task will take a long time [4]. Thus it is rather challenging to support macro-tasks, including automatically splitting a macro-task, assigning tasks to crowd or machines, and automatically aggregating the answers.

8.2.5 Privacy

There are several types of privacy issued in crowdsourcing. First, the requester wants to protect the privacy of their tasks [7, 8]. The tasks may contain sensitive attributes and could cause privacy leakage. Malicious workers could link them with other public datasets to reveal individual private information. Although the requester can publish anonymity data to the workers using existing privacy techniques, e.g., K-Anonymity, it may lower down the quality as the workers cannot get the precise data. Thus it is challenging to trade off the accuracy and privacy for requesters. Second, the workers have privacy-preserving requirement. Personal information of workers can be inferred from the answers provided by the workers, such as their location, profession, and hobby. On the other hand, the requester wants to assign their tasks to appropriate workers that are skilled at their tasks (or close to the tasks). And it is challenging to devise privacy-preserving task assignment techniques.

8.2.6 Mobile Crowdsourcing

With the growing popularity of smartphones, there are emerging more and more mobile crowdsourcing platforms, e.g., gMission [3], Waze [2], and ChinaCrowd [1]. These mobile platforms pose many new challenges for crowdsourced data management. First, many more factors (e.g., spatial distance, mobile user interface) will affect workers' latency and quality. It is more challenging to control quality, latency, and cost for mobile platforms. Second, traditional crowdsourcing platforms adopt worker selection model to assign tasks; however mobile crowdsourcing requires to support server assignment model (see Sect. 7.11). It calls for new task assignment techniques.

References

1. Chinacrowd. http://www.chinacrowds.com
2. Waze. https://www.waze.com
3. Chen, Z., Fu, R., Zhao, Z., Liu, Z., Xia, L., Chen, L., Cheng, P., Cao, C.C., Tong, Y., Zhang, C.J.: gmission: a general spatial crowdsourcing platform. PVLDB 7(13), 1629–1632 (2014)
4. Haas, D., Ansel, J., Gu, L., Marcus, A.: Argonaut: Macrotask crowdsourcing for complex data processing. PVLDB 8(12), 1642–1653 (2015)

5. Mozafari, B., Sarkar, P., Franklin, M., Jordan, M., Madden, S.: Scaling up crowd-sourcing to very large datasets: a case for active learning. PVLDB **8**(2), 125–136 (2014)
6. Wang, J., Krishnan, S., Franklin, M.J., Goldberg, K., Kraska, T., Milo, T.: A sample-and-clean framework for fast and accurate query processing on dirty data. In: SIGMOD, pp. 469–480 (2014)
7. Wu, S., Wang, X., Wang, S., Zhang, Z., Tung, A.K.H.: K-anonymity for crowdsourcing database. TKDE **26**(9), 2207–2221 (2014)
8. Yuan, D., Li, G., Li, Q., Zheng, Y.: Sybil defense in crowdsourcing platforms. In: CIKM, pp. 1529–1538 (2017)

Glossary

AMT	Amazon Mechanical Turk.
HIT	Human Intelligence Task.
Requester	The user that publishes machine-hard tasks on crowdsourcing platforms.
Worker	The user that answers tasks on crowdsourcing platforms.
Platform	Manage the tasks and assign tasks to workers.
Quality Control	The techniques that improve the quality of tasks.
Cost Control	The techniques that reduce the cost of tasks.
Latency Control	The techniques that reduce the latency of tasks.
Golden Tasks	Tasks with ground truth.
Qualification Test	Testing workers using golden tasks before they answer normal tasks.
Hidden Test	Testing workers using golden tasks when they answer normal tasks.
Truth Inference	Inferring the truth of a task.
Worker Modeling	Modeling workers' quality.
Task Modeling	Modeling tasks' difficulty or domains.
Task Assignment	Assigning task(s) to a worker.
Answer Aggregation	Aggregating the truth of a task based on the answers from different workers.
Task Selection	Selecting a set of tasks to assign.
Task Design	Designing task UIs.
Task Pruning	Pruning the tasks that are not necessary to ask workers.
Answer Deduction	Deducing the answers of a set of tasks based on those of other tasks.
Latency Modeling	Modeling task latency.

© Springer Nature Singapore Pte Ltd. 2018
G. Li et al., *Crowdsourced Data Management*,
https://doi.org/10.1007/978-981-10-7847-7

Printed in the United States
By Bookmasters